SUPERSYMMETRY, SUPERFIELDS AND SUPERGRAVITY:

AN INTRODUCTION

Graduate Student Series in Physics
Other books in the series

Colour Centres and Imperfections in Insulators and Semiconductors
P D TOWNSEND and J C KELLY

The Superconducting State
A D C GRASSIE

Refrigeration and Thermometry below One Kelvin
D S BETTS

Plasma Physics
E W LAING

Weak Interactions
D BAILIN

Gauge Theories in Particle Physics
I J R AITCHISON and A J G HEY

Collective Effects in Solids and Liquids
N H MARCH and M PARRINELLO

Hadron Interactions
P D B COLLINS and A D MARTIN

Superfluidity and Superconductivity
D R TILLEY and J TILLEY

Introduction to Gauge Field Theory
D BAILIN and A LOVE

Mechanics of Deformable Media
A B BHATIA and R N SINGH

GRADUATE STUDENT SERIES IN PHYSICS

Series Editor: Professor Douglas F Brewer, M.A., D.Phil.
Professor of Experimental Physics, University of Sussex

SUPERSYMMETRY, SUPERFIELDS AND SUPERGRAVITY:

AN INTRODUCTION

PREM P SRIVASTAVA
Centro Brasileiro de Pesquisas Físicas
Rio de Janeiro, Brazil

ADAM HILGER, BRISTOL AND BOSTON
Published in association with the University of Sussex Press

British Library Cataloguing in Publication Data

Srivastava, Prem P.
Supersymmetry, superfields and supergravity: an introduction. —— (Graduate student series in physics)
1. Supersymmetry
I. Title II. Series
530.1′43 QC174.17.S9

ISBN 0-85274-571-0
ISBN 0-85274-575-3 Pbk

Published under the Adam Hilger imprint by IOP Publishing Ltd
Techno House, Redcliffe Way, Bristol BS1 6NX, England
PO Box 230, Accord, MA 02018 USA
in association with the University of Sussex Press

Typeset by Mathematical Composition Setters Ltd, Salisbury, UK

Printed in Great Britain by J W Arrowsmith Ltd, Bristol

PREFACE

The present book aims at introducing to graduate students the subject of supersymmetry in a self-contained text. The basic techniques are explained in sufficient detail to allow the student to pursue with some confidence more advanced literature and some lines of current research work. The book grew out of graduate courses delivered in 1983 at CBPF, Rio de Janeiro and at Instituto de Física Teórica, São Paulo.

In Chapter 1, after a detailed discussion of the notation, the algebras of supersymmetry and R-symmetry generators are motivated by considering the simplest field theory model. Chapter 2 deals with its realisations on one-particle states and on a supermultiplet of component fields. The Wess–Zumino model is described in detail, including a discussion on the realisation of R-symmetry and supermultiplets of currents and anomalies. The realisation of the algebra on superspace and superfields forms the substance of Chapters 3–5 and Chapter 7 where the superfields are used to construct the general supersymmetric and gauge invariant action for the Yang–Mills theory in interaction with the matter. Chapter 6 on spontaneously broken supersymmetry is inserted before Chapter 7 on the non-Abelian supersymmetric gauge theory in order to sustain the enthusiasm of the reader with the important physical implications of supersymmetry in a field theory. Chapter 8 discusses integration over Grassmann variables, the rules of integration by parts over the whole of superspace and other technical details needed for the next two chapters. The superfield propagators are derived in Chapter 9 as the Green functions of the corresponding equations of motion. The introduction of unconstrained superfield potentials for the constrained chiral superfields with the corresponding inclusion of a gauge-fixing term to the action is emphasised. Chapter 10 illustrates the compactness and the power of superfield perturbation theory through some detailed calculations of the radiative corrections for the case of a chiral superfield as well as for the gauge superfield in the Abelian case. A brief discussion on the non-renormalisation theorem and the superfield tadpole method for calculating the effective potential is also included. The last Chapter deals with local supersymmetry and the supergravity Lagrangian. An outline of gravity induced supersymmetry breaking and the super-Higgs effect is given and illustrated through several models including the recent very promising 'no scale' models.

The book owes much to the opportunity for several short visits I had at the International Centre for Theoretical Physics, Trieste, under the associate scheme which allowed me to keep informed of the latest developments and build an overview. For this my grateful acknowledgements are due to Professor Abdus Salam, IAEA, UNESCO and CNPq as well as CAPES of Brasil. The hospitality extended to me at CERN Theory Division during several visits under a CNPq–CERN agreement is also thankfully acknowledged. I also acknowledge my debt to Dr David Bailin for his initial encouragement and suggestions, a critical reading and constructive suggestions during the writing of this book. I would like to thank also Drs Abraham Zimerman and Paulo Leal Ferreira for the warm hospitality extended to me at IFT and for their encouragement (rather, the insistence of the former) to write the initial lecture notes.

Prem P Srivastava December 1985

CONTENTS

1

SUPERSYMMETRY

1.1 Introduction

Supersymmetry is a relativistic symmetry between bosons and fermions. Is the only known way available at the present to unify the four-dimensional space–time and internal symmetries of the *S*-matrix in relativistic particle theory. It unites in a single supermultiplet particles with different intrinsic spins differing by units of 1/2 so that fermions become superpartners of bosons and vice versa. The symmetry generators are consequently fermionic and obey anticommutation relations. In the past such attempts at higher symmetries were concerned with multiplets carrying particles with the same statistics though different spins and the symmetry generators were bosonic and generated the ordinary Lie algebra. Several no-go theorems were discovered to show that no non-trivial relativistic unification was possible with such generators. The graded rather than the ordinary Lie-algebraic structure of the superalgebra allows us to avoid these theorems. Another striking feature of supersymmetric field theories is that they are less ultraviolet divergent than the corresponding non-supersymmetric theory, due to the miraculous cancellations of the divergences of fermionic loops with those of the bosonic loops. Models with more than one supersymmetry exhibit even more ultraviolet convergence; for example, $N=4$ supersymmetric Yang–Mills theory has been shown recently to be a finite theory. When supersymmetry is raised to a local supersymmetry the introduction of gravitation is automatic and leads to the possibility of unifying gravitational with the strong, weak and electromagnetic interactions.

The first examples of Bose–Fermi symmetry appeared in the Neveu and Schwarz (1971) and Ramond (1971) spinning string model with particles of half-integer spins, and in the extension of the Poincaré algebra by Gol'fand and Likhtman (1971) followed by the nonlinear realisation of the supersymmetry algebra by Volkov and Akulov (1973). The introduction of linear representations of supersymmetry in the context of quantum field theory was first given by Wess and Zumino (1974b; also Zumino 1975). Soon after, Salam and Strathdee (1974b) and Ferrara *et al.* (1974) found the realisation of supersymmetry generators on a superspace of coordinates and introduced superfields over it to describe a supersymmetry multiplet. This lead to a rapid development in finding supersymmetric extensions of ordinary field theories and in the study of their miraculous properties.

1.2 Dirac and Weyl spinors

1.2.1 Dirac spinor representation of the Lorentz group

We shall work in four-dimensional Minkowski space with flat metric $\eta_{lm} = \text{diag}(-1, 1, 1, 1)$ where $l = 0, 1, 2, 3$ denotes the space–time indices. The restricted homogeneous Lorentz group (Lorentz group) of transformations are induced on space–time coordinates, $x'^{\,l} = \Lambda^l{}_m x^m$, by the group SO(1,3) of real 4×4 matrices $\mathbf{\Lambda} = (\Lambda^l{}_m)$ satisfying $\eta_{pq}\Lambda^p{}_l\Lambda^q{}_m = \eta_{lm}$ which simply expresses the equivalence of $\{\mathbf{\Lambda}\}$ to the corresponding contragradient representation $\{\mathbf{\Lambda}^{-\text{IT}}\}$ written as $\{\mathbf{\Lambda}^{-\text{IT}}\} \sim \{\mathbf{\Lambda}\}$. The infinitesimal generators † Σ_{lm} are easily found to be $(\Sigma^{lm})^p{}_q = -\text{i}(\eta^{lp}\delta^m_q - \eta^{mp}\delta^l{}_q)$. They span a Lie algebra under the commutator which is found to be

$$[M_{lm}, M_{pq}] = \text{i}(\eta_{lp}M_{mq} - \eta_{mp}M_{lq} + \eta_{mq}M_{lp} - \eta_{lq}M_{mp}) \tag{1.1}$$

where for generality we have written $M_{lm} = -M_{ml}$ in place of Σ_{lm}. The exponential parametrisation of $\mathbf{\Lambda}$ may be written as

$$\mathbf{\Lambda} = \exp\left(\frac{\text{i}}{2}\lambda_{lm}\Sigma^{lm}\right)$$

where $\lambda_{lm} = -\lambda_{ml}$ are six (essential) parameters characterising a Lorentz rotation. We find that $(J_1, J_2, J_3) = (M_{23}, M_{31}, M_{12})$ satisfy the algebra of angular momentum operators and generate space rotations while M_{01}, M_{02}, M_{03} are Lorentz boosts. The algebra (1.1) which is valid in any representation may alternatively be written in terms of six generators $J_\pm$ with $J^i_\pm = \frac{1}{2}(M^{jk} \pm \text{i}M^{0i})$, ijk cyclic $1 \to 2 \to 3 \to 1$, and

$$[J^i_\pm, J^j_\pm] = \text{i}\varepsilon^{ijk} J_\pm{}^k \qquad [J^i_+, J^j_-] = 0 \tag{1.2}$$

where i, j, k denote space indices 1, 2, 3. The finite dimensional (in general non-unitary) irreducible representations $D(j_+, j_-)$ may thus be classified‡ by two half-integer numbers corresponding to the eigenvalues $j_\pm(j_\pm + 1)$ of the two Casimir operators $\mathbf{J}^2_\pm$. The representations $D(\frac{1}{2}, 0)$ and $D(0, \frac{1}{2})$ correspond to Weyl 2-spinors which will be extensively used in our discussions.

A realisation of (1.1) over space–time coordinates is given by the generalised angular momentum differential operators $L_{lm} = (1/\text{i})(x_l\partial_m - x_m\partial_l)$. We verify that $\delta x^l = \lambda^l{}_m x^m = -(\text{i}/2)(\lambda_{pq}L^{pq})x^l$.

A different realisation of the Lorentz algebra may be obtained in terms of the element $\gamma^0, \gamma^1, \gamma^2, \gamma^3$ of a Clifford algebra over four-dimensional

† det $\Lambda = 1$, sgn $\Lambda^0{}_0 = +$ and for infinitesimal transformations $\Lambda^l{}_m = \delta^l_m + \lambda^l_m + \ldots = \delta^l_m + (\text{i}/2)\lambda_{pq}(\Sigma^{pq})^l{}_m + \ldots$

‡ See for general discussion Naimark (1957); see also Heine (1957).

Minkowski space which satisfy

$$\{\gamma^l, \gamma^m\} = 2\eta^{lm}\mathbf{I}. \tag{1.3}$$

We verify that $M_{lm} = (1/4\mathrm{i})[\gamma_l, \gamma_m]$ do satisfy (1.1) in view of (1.3). The Lie group elements $\mathbf{S}(\Lambda)$ are then parametrised as

$$\mathbf{S}(\Lambda) = \exp\frac{\mathrm{i}}{2}\lambda_{lm}M^{lm} = \exp\frac{1}{8}\lambda_{lm}[\gamma^l, \gamma^m]. \tag{1.4}$$

We verify that

$$[M^{lm}, \gamma^n] = -(\Sigma^{lm})^n{}_p\gamma^p = \mathrm{i}(\eta^{ln}\gamma^n - \eta^{mn}\gamma^l)$$

$$\gamma^l = \Lambda^l{}_m\mathbf{S}(\Lambda)\gamma^m\mathbf{S}^{-1}(\Lambda) \tag{1.5}$$

which show that γ^l are invariant under Lorentz rotations and l is a four-vector index.

An irreducible representation of γ^l may be obtained in terms of 4×4 complex matrices. The representation matrices $\mathbf{S}(\Lambda)$ then act on a complex spinor space spanned by the Dirac 4-spinors $\psi(x)$:

$$\psi(x) = \begin{pmatrix}\psi_1\\ \psi_2\\ \psi_3\\ \psi_4\end{pmatrix}(x) \xrightarrow{\Lambda} \psi'(x') = \mathbf{S}(\Lambda)\psi(x)$$

$$\det\mathbf{S}(\Lambda) = 1. \tag{1.6}$$

The representations $\{\mathbf{S}^{-\mathrm{IT}}\}$, $\{\mathbf{S}^*\}$ and $\{\mathbf{S}^{\dagger-1}\}$ are all equivalent to $\{\mathbf{S}\}$ because of the fundamental lemma which ensures the equivalence† of each of the set of gamma matrices $\{\gamma^{*l}\}$,$\{-\gamma^{l\mathrm{T}}\}$, $\{\gamma^{l\dagger}\}$ etc, also satisfying (1.3), to the set $\{\gamma^l\}$. Henceforth, we will work in a representation such that

$$\gamma^0\gamma^l\gamma^0 = \gamma^{l\dagger} \qquad \gamma^0\gamma_5\gamma^0 = -\gamma_5^\dagger = \gamma_5 \tag{1.7}$$

where $\gamma_5 = \gamma^0\gamma^1\gamma^2\gamma^3$, $\gamma_5^2 = -\mathbf{I}$. The adjoint defined by $\bar\psi = \psi^\dagger\gamma^0$ is seen to transform as

$$\bar\psi \xrightarrow{\Lambda} \bar\psi\mathbf{S}^{-1}(\Lambda).$$

Owing to the existence of the charge conjugation matrix $\mathbf{C}$ satisfying

$$\mathbf{C}^{-1}\gamma^l\mathbf{C} = -\gamma^{l\mathrm{T}} \tag{1.8}$$

we may also define a charge conjugate spinor $\psi_c = \mathbf{C}\bar\psi^{\mathrm{T}} = \mathbf{C}\gamma^{0\mathrm{T}}\psi^* = \mathbf{D}\psi^*$. From $\mathbf{C}^{-1}\mathbf{S}(\Lambda)\mathbf{C} = \mathbf{S}^{-\mathrm{IT}}(\Lambda)$ it follows that $\psi_c \to \mathbf{S}(\Lambda)\psi_c$ like ψ. In four space–time dimensions $\mathbf{C}$ is necessarily antisymmetric, $\mathbf{C}^{\mathrm{T}} = -\mathbf{C}$ in order

† For example, there exist non-singular matrices such that $\mathbf{A}\gamma^l\mathbf{A}^{-1} = -\gamma^{l\dagger}$, $\mathbf{A}\mathbf{S}(\Lambda)\mathbf{A}^{-1} = \mathbf{S}^{\dagger-1}(\Lambda)$, $\mathbf{C}^{-1}\gamma^l\mathbf{C} = -\gamma^{l\mathrm{T}}$. This is true for any space–time dimension. See, for example Schweber (1961).

that we obtain ten symmetric $\gamma^l \mathbf{C}$, $[\gamma^l, \gamma^m]\mathbf{C}$ and six antisymmetric $\mathbf{C}, \gamma_5\mathbf{C}, \gamma_5\gamma^l\mathbf{C}$ linearly independent matrices. We may fix $\mathbf{C}$, apart from a sign, by imposing, say†, $(\psi_c)_c = \psi$ which requires $\mathbf{D}^*\mathbf{D} = 1$ or $\mathbf{CC}^* = -1$. Consequently, we find $\bar{\psi}_c = \psi^T\mathbf{C}^* = -\psi^T\mathbf{C}^{-1}$ and $\bar{\psi}_c\mathbf{M}\psi_c = -\psi^T\mathbf{C}^{-1}\mathbf{MC}\bar{\psi}^T = -\bar{\psi}(\mathbf{MC})^T\mathbf{C}^{-1}\psi$. In particular‡, $\bar{\psi}_c\psi_c = \bar{\psi}\psi$, $\bar{\psi}_c\gamma^l\psi_c = -\bar{\psi}\gamma^l\psi$ justify the name charge conjugate.

We may also define in four space–time dimensions a Majorana spinor with half as many independent components by imposing the Lorentz invariant Majorana condition $\psi_c = \psi$. The constraint is only consistent if $(\psi_c)_c = \psi$ which may not be possible in other space–time dimensions. The vector and tensor currents vanish identically for Majorana spinors and the Majorana condition is incompatible with the U(1) gauge transformation $\psi \to e^{i\alpha}\psi$, α real. However, the following axial or chiral U(1) group gauge transformation may be consistently defined even for Majorana spinors:

$$\begin{aligned} \psi'(x) &= e^{\alpha\gamma_5}\psi(x) \qquad & \bar{\psi}'(x) &= \bar{\psi}(x)e^{\alpha\gamma_5} \\ \psi_c'(x) &= e^{\alpha\gamma_5}\psi_c \qquad & \bar{\psi}_c'(x) &= \bar{\psi}_c(x)e^{\alpha\gamma_5}. \end{aligned} \tag{1.9}$$

We find that the scalar, pseudoscalar and tensor bilinears of ψ are not chiral invariant while the axial current and vector current (hence the kinetic term for α = constant) are left invariant. In chiral invariant theory the corresponding conserved Noether current contains an axial vector term $\bar{\psi}\gamma_5\gamma^l\psi$.

A couple of Majorana spinors ξ, η define a Dirac spinor ψ as follows

$$\psi = \xi + i\eta \qquad \psi_c = \xi - i\eta \tag{1.10}$$

where

$$\xi = \tfrac{1}{2}(\psi + \psi_c) \qquad \eta = \frac{1}{2i}(\psi - \psi_c). \tag{1.11}$$

The ordinary U(1) gauge transformation is thus well defined for a couple of Majorana spinors.

The left and right chiral projections ψ_L, ψ_R

$$\psi_{\substack{L\\R}} = \tfrac{1}{2}(1 \mp i\gamma_5)\psi \qquad \psi^{\dagger}_{\substack{L\\R}} = \psi^{\dagger}\tfrac{1}{2}(1 \mp i\gamma_5) \qquad \bar{\psi}_{\substack{L\\R}} = \bar{\psi}\tfrac{1}{2}(1 \pm i\gamma_5)$$

$$\psi = \psi_L + \psi_R \qquad \bar{\psi} = \bar{\psi}_L + \bar{\psi}_R \tag{1.12}$$

carry separately an irreducible representation of the homogeneous Lorentz group, $\psi_L \to \mathbf{S}(\Lambda)\psi_L$, $\psi_R \to \mathbf{S}(\Lambda)\psi_R$. For massless spin-1/2 particles the γ_5-chirality and helicity stand in one-to-one correspondence and left-handed stands for the 'spin' pointing along the direction opposite to that

† This may not be possible in other space–time dimensions even though ψ_c may be defined.

‡ We also note $\bar{\psi}_c\gamma_5\gamma^l\psi_c = \bar{\psi}\gamma_5\gamma^l\psi$ and for the kinetic term $\bar{\psi}_c\gamma^l\partial_l\psi_c - (\partial_l\bar{\psi}_c)\gamma^l\psi_c = \bar{\psi}\gamma^l\partial_l\psi - (\partial_l\bar{\psi})\gamma^l\psi$.

of momentum. The electron neutrino, for example, is left-handed. In grand unified theories ψ_L, ψ_R may even belong to different representations of the internal symmetry gauge group. Under the chiral gauge transformation (1.9)

$$\psi_L'(x) = e^{i\alpha}\psi_L \qquad \psi_R'(x) = e^{-i\alpha}\psi_R(x) \tag{1.13}$$

so that ψ_L and ψ_R carry $U(1)_A$ charges with opposite sign. We also note that

$$\begin{aligned} \bar\psi M\psi &= \bar\psi_L M\psi_R + \bar\psi_R M\psi_L \qquad &\text{for } M = 1, \gamma_5, \gamma^l\gamma^m \\ &= \bar\psi_L M\psi_L + \bar\psi_R M\psi_R \qquad &\text{for } M = \gamma^l, \gamma_5\gamma^l \end{aligned} \tag{1.14}$$

and for Majorana spinors $\bar\psi_L\gamma^l\psi_L = -\bar\psi_R\gamma^l\psi_R$, $\bar\psi_L\gamma^l\gamma^m\psi_R = -\bar\psi_R\gamma^l\gamma^m\psi_L$. The Majorana condition $\psi_c = \psi$ written in terms of chiral spinors reads as

$$\psi_{L(R)} = \mathbf{C}\bar\psi^T_{R(L)} \qquad \bar\psi_{L(R)} = -\psi^T_{R(L)}\mathbf{C}^{-1} \qquad \text{(Majorana spinor).} \tag{1.15}$$

When the discrete transformations parity, time inversion and space–time inversion are also included, the spinor representation is not reducible. For example, the parity operation† may be defined by

$$\psi'(x') = \psi'(x^0, -\boldsymbol{x}) = -ia\gamma^0\psi(x) \tag{1.16}$$

where in view of the double-valuedness of the spinor representation $a^2 = \pm 1$. For Majorana spinors we must have $a = \pm i$. In terms of chiral spinors

$$\psi_L'(x') = -ia\gamma^0\psi_R(x) \qquad \psi_R'(x') = -ia\gamma^0\psi_L(x) \tag{1.17}$$

and we need both ψ_L and ψ_R to realise parity.

1.2.2 Two-component Weyl spinors

In the Weyl representation of gamma matrices in terms of a set of odd matrices

$$\gamma^l = i\begin{pmatrix} \mathbf{0} & \sigma^l \\ \bar\sigma^l & \mathbf{0} \end{pmatrix} \tag{1.18}$$

where $\sigma^l, \bar\sigma^l$ are 2×2 matrices $\mathbf{S}(\Lambda)$ takes a block diagonal form. We may then work in terms of the 2-spinor formulation. The Fierz rearrangement of fermionic fields becomes very simple and complicated Bianchi identities may be handled in a straightforward fashion.

From (1.3) we require

$$\sigma^l\bar\sigma^m + \sigma^m\bar\sigma^l = \bar\sigma^l\sigma^m + \bar\sigma^m\sigma^l = -2\eta^{lm}. \tag{1.19}$$

† For space reflections we find from (1.5) that $\gamma^0 = \mathbf{S}\gamma^0\mathbf{S}^{-1}$, $\gamma^i = -\mathbf{S}\gamma^i\mathbf{S}^{-1}$ and we may choose $\mathbf{S} = -ia\gamma^0$. For other cases see, for example, Marshak *et al.* (1969).

Taking into consideration (1.7) and making a convenient choice of phase factor in γ^0 we may take

$$\sigma^l = (\mathbf{I}, \boldsymbol{\sigma}) \qquad \bar{\sigma}^l = (\mathbf{I}, -\boldsymbol{\sigma}) \tag{1.20}$$

where $\boldsymbol{\sigma}$ are the Pauli matrices. We then find

$$\mathrm{i}\gamma_5 = \begin{pmatrix} -\mathbf{I} & \mathbf{0} \\ \mathbf{0} & \mathbf{I} \end{pmatrix} \qquad \mathbf{C} = -\gamma^0\gamma^2 = \begin{pmatrix} -\sigma_2 & \mathbf{0} \\ \mathbf{0} & \sigma_2 \end{pmatrix} = -\mathbf{C}^{\mathrm{T}} = -\mathbf{C}^{*-1} \tag{1.21}$$

where a phase factor in $\mathbf{C}$ has been chosen conveniently. For $\mathbf{S}(\Lambda)$ we obtain

$$\mathbf{S}(\Lambda) = \begin{pmatrix} \mathbf{S}_1(\Lambda) & \mathbf{0} \\ \mathbf{0} & \mathbf{S}_1^{-1\dagger}(\Lambda) \end{pmatrix} \tag{1.22}$$

where

$$\mathbf{S}_1(\Lambda) = \exp(-\tfrac{1}{2}\lambda_{lm}\sigma^{lm}) \qquad \mathbf{S}_1^{\dagger-1}(\Lambda) = \exp(-\tfrac{1}{2}\lambda_{lm}\bar{\sigma}^{lm})$$
$$\det \mathbf{S}_1 = \det \mathbf{S}_1^{\dagger-1} = 1 \tag{1.23}$$

and

$$\sigma^{lm} = \tfrac{1}{4}(\sigma^l\bar{\sigma}^m - \sigma^m\bar{\sigma}^l) = \{-\tfrac{1}{2}\boldsymbol{\sigma}, -\tfrac{1}{2}\mathrm{i}\boldsymbol{\sigma}\}$$
$$\bar{\sigma}^{lm} = \tfrac{1}{4}(\bar{\sigma}^l\sigma^m - \bar{\sigma}^m\sigma^l) = \{\tfrac{1}{2}\boldsymbol{\sigma}, -\tfrac{1}{2}\mathrm{i}\boldsymbol{\sigma}\}. \tag{1.24}$$

The generators $\mathrm{i}\sigma^{lm}$ and $\mathrm{i}\bar{\sigma}^{lm}$ individually satisfy (1.1) and $\mathbf{S}_1$ and $\mathbf{S}_1^{-1\dagger}$ furnish two inequivalent (2×2) matrix representations $\mathbf{D}(\tfrac{1}{2}, 0)$ and $\mathbf{D}(0, \tfrac{1}{2})$ respectively of the SL(2, C) group which is the universal covering group of SO(1, 3). The SL(2, C) spinors over which they act will be indicated by χ and $\bar{\chi}$

$$\chi'(x') = \mathbf{S}_1(\Lambda)\chi(x) \qquad \bar{\chi}'(x') = \mathbf{S}_1^{-1\dagger}(\Lambda)\bar{\chi}(x). \tag{1.25}$$

We derive from $\sigma_2\boldsymbol{\sigma}\sigma_2 = -\boldsymbol{\sigma}^{\mathrm{T}} = -\boldsymbol{\sigma}^*$ that

$$\bar{\sigma}^l = \sigma_2\sigma^{l*}\sigma_2 = \sigma_2\sigma^{l\mathrm{T}}\sigma_2 \tag{1.26}$$

and, consequently,

$$\mathbf{S}_1 = \sigma_2\mathbf{S}_1^{-1\mathrm{T}}\sigma_2 \qquad \mathbf{S}_1^* = \sigma_2\mathbf{S}_1^{-1\dagger}\sigma_2 \tag{1.27}$$

which shows the equivalence of the representations $\{\mathbf{S}_1^{-1\mathrm{T}}\}$ to $\{\mathbf{S}_1\}$ and $\{\mathbf{S}_1^*\}$ to $\{\mathrm{S}_1^{-1\dagger}\}$.

The Dirac spinor now assumes the form

$$\psi = \begin{pmatrix} \chi \\ \bar{\chi}_{\mathrm{c}} \end{pmatrix} \qquad \bar{\psi} = \mathrm{i}(\bar{\chi}_{\mathrm{c}}^{\dagger}, \chi^{\dagger})$$

while its chiral projections are

$$\psi_{\mathrm{L}} = \begin{pmatrix} \chi \\ 0 \end{pmatrix} \qquad \psi_{\mathrm{R}} = \begin{pmatrix} 0 \\ \bar{\chi}_{\mathrm{c}} \end{pmatrix}.$$

The charge conjugate spinor is

$$\psi_c = -\gamma^2\psi^* = \begin{pmatrix} -i\sigma_2\bar{\chi}_c^* \\ i\sigma_2\chi^* \end{pmatrix}$$

and the Majorana condition reads as $\chi = -i\sigma_2\bar{\chi}_c^*$. It is elegant to introduce 2-spinor indices

$$\chi = (\chi_2) = \begin{pmatrix}\chi_1 \\ \chi_2\end{pmatrix} \qquad \bar{\chi} = \begin{pmatrix}\bar{\chi}^{\dot{1}} \\ \bar{\chi}^{\dot{2}}\end{pmatrix}. \tag{1.28}$$

Now from (1.25) and (1.27) it is clear that $(\sigma_2\chi)$ transforms according to $\mathbf{S}_{\mathrm{I}}^{-1\mathrm{T}}$ while $(\sigma_2\bar{\chi})$ transforms according to $\mathbf{S}_{\mathrm{I}}^*$, and $(\sigma_2\chi)^{\mathrm{T}}\eta$ as well as $(\sigma_2\bar{\chi})^{\mathrm{T}}\bar{\eta}$ are SL(2, C) or Lorentz invariants. We may thus define

$$(\chi^\alpha) = \begin{pmatrix}\chi^1 \\ \chi^2\end{pmatrix} = i\sigma_2\chi \qquad (\bar{\chi}_{\dot{\alpha}}) = \begin{pmatrix}\bar{\chi}_{\dot{1}} \\ \bar{\chi}_{\dot{2}}\end{pmatrix} = -i\sigma_2\bar{\chi} \tag{1.29}$$

and write the invariants as $\chi^\alpha\eta_\alpha$ and $\bar{\chi}_{\dot{\alpha}}\bar{\eta}^{\dot{\alpha}}$, i.e. contract an upper index with a lower index of the same type. Introducing metric tensors to raise and lower the indices

$$\begin{aligned} &(\varepsilon^{\alpha\beta}) = (\varepsilon^{\dot{\alpha}\dot{\beta}}) = i\sigma_2 \qquad (\varepsilon_{\alpha\beta}) = (\varepsilon_{\dot{\alpha}\dot{\beta}}) = -i\sigma_2 \\ &\varepsilon_{\alpha\beta}\varepsilon^{\beta\sigma} = \delta^\sigma_\alpha \qquad \varepsilon^{\dot{\alpha}\dot{\beta}}\varepsilon_{\dot{\beta}\dot{\sigma}} = \delta^{\dot{\alpha}}_{\dot{\sigma}} \end{aligned} \tag{1.30}$$

we find $\chi^\alpha = \varepsilon^{\alpha\beta}\chi_\beta$, $\bar{\chi}_{\dot{\alpha}} = \varepsilon_{\dot{\alpha}\dot{\beta}}\bar{\chi}^{\dot{\beta}}$ etc. If we write $\mathbf{S}_{\mathrm{I}} = (\mathbf{S}_{\mathrm{I}\alpha}{}^\beta)$ and $\mathbf{S}_{\mathrm{I}}^* = (\mathbf{S}^*_{\mathrm{I}\dot{\alpha}}{}^{\dot{\beta}})$ we find $\varepsilon^{\alpha\beta}\mathbf{S}_{\mathrm{I}\rho}{}^\sigma\varepsilon_{\sigma\beta} = (\sigma_2\mathbf{S}_{\mathrm{I}}\sigma_2)^\alpha{}_\beta = (\mathbf{S}_{\mathrm{I}}^{-1\mathrm{T}})^\alpha{}_\beta$ and $\varepsilon^{\dot{\alpha}\dot{\rho}}\mathbf{S}^*_{\mathrm{I}\dot{\rho}}{}^{\dot{\sigma}}\varepsilon_{\dot{\sigma}\dot{\beta}} = (\sigma_2\mathbf{S}_{\mathrm{I}}^*\sigma_2)^{\dot{\alpha}}{}_{\dot{\beta}} = (\mathbf{S}_{\mathrm{I}}^{-1\dagger})^{\dot{\alpha}}{}_{\dot{\beta}}$. Consequently,

$$\begin{aligned} &\chi'_\alpha = \mathbf{S}_{\mathrm{I}\alpha}{}^\beta\chi_\beta \qquad \chi'^\alpha = \varepsilon^{\alpha\beta}\chi'_\beta = \varepsilon^{\alpha\rho}\mathbf{S}_{\mathrm{I}\rho}{}^\sigma\varepsilon_{\sigma\beta}\chi^\beta = (\mathbf{S}_{\mathrm{I}}^{-1\mathrm{T}})^\alpha{}_\beta\chi^\beta \\ &\bar{\chi}'_{\dot{\alpha}} = \mathbf{S}^*_{\mathrm{I}\dot{\alpha}}{}^{\dot{\beta}}\bar{\chi}_{\dot{\beta}} \qquad \bar{\chi}'^{\dot{\alpha}} = (\mathbf{S}_{\mathrm{I}}^{-1\dagger})^{\dot{\alpha}}{}_{\dot{\beta}}\bar{\chi}^{\dot{\beta}}. \end{aligned} \tag{1.31}$$

We will henceforth identify $\bar{\chi}_{\dot{\alpha}}$ with the complex conjugate χ_α^* and $\bar{\chi}_{\dot{\alpha}}^*$ with χ_α. It is easily shown that the metric tensors as well as δ^β_α are invariant tensors.

Other useful SL(2, C) invariant tensors may be obtained from (1.5) which reads as

$$\sigma^l = \Lambda^l{}_m\mathbf{S}_{\mathrm{I}}\sigma^m\mathbf{S}_{\mathrm{I}}^\dagger \qquad \bar{\sigma}^l = \Lambda^l{}_m\mathbf{S}_{\mathrm{I}}^{\dagger-1}\bar{\sigma}^m\mathbf{S}_{\mathrm{I}}^{-1}. \tag{1.32}$$

With the indices defined in (1.31) we obtain the indices of σ^l and $\bar{\sigma}^l$:

$$\sigma^l = (\sigma^l_{\alpha\dot{\beta}}) \qquad \bar{\sigma}^l = (\bar{\sigma}^{l\dot{\alpha}\beta}) \tag{1.33}$$

and

$$\begin{aligned} \sigma^l_{\alpha\dot{\beta}} &= \Lambda^l{}_m\mathbf{S}_{\mathrm{I}\alpha}{}^\rho\sigma^m_{\rho\dot{\sigma}}(\mathbf{S}_{\mathrm{I}}^*)_{\dot{\beta}}{}^{\dot{\sigma}} \\ \bar{\sigma}^{l\dot{\alpha}\beta} &= \Lambda^l{}_m(\mathbf{S}_{\mathrm{I}}^{-1\dagger})^{\dot{\alpha}}{}_{\dot{\rho}}\bar{\sigma}^{m\dot{\rho}\sigma}\mathbf{S}_{\mathrm{I}}^{-1}{}_\sigma{}^\beta \\ &= \Lambda^l{}_m(\mathbf{S}_{\mathrm{I}}^{-1\dagger})^{\dot{\alpha}}{}_{\dot{\rho}}(\mathbf{S}_{\mathrm{I}}^{-1\mathrm{T}})^\beta{}_\sigma\bar{\sigma}^{m\dot{\rho}\sigma} \end{aligned} \tag{1.34}$$

explicitly show their invariant nature. The following hermiticity property

$$\bar{\sigma}^{l}_{\dot{\alpha}\beta} = \sigma^{l}_{\beta\dot{\alpha}} \tag{1.35}$$

follows from

$$\bar{\sigma}^{l}_{\dot{\alpha}\beta} = \varepsilon_{\dot{\alpha}\dot{\sigma}}\varepsilon_{\beta\rho}\bar{\sigma}^{l\dot{\sigma}\rho} = (\sigma_2\bar{\sigma}^{l}\sigma_2)_{\dot{\alpha}\beta} = (\sigma^{l\mathrm{T}})_{\dot{\alpha}\beta} = \sigma^{l}_{\beta\dot{\alpha}} \tag{1.36}$$

and $\sigma^{l\mathrm{T}} = \sigma^{l*}$ further leads to $\bar{\sigma}^{l}_{\dot{\alpha}\beta} = (\sigma^{l}_{\alpha\dot{\beta}})^{*}$.

The *completeness relations* involving σ^{l}, $\bar{\sigma}^{m}$ may be derived easily. The first one is

$$\mathrm{Tr}(\sigma^{l}\bar{\sigma}^{m}) = \sigma^{l}_{\alpha\dot{\beta}}\bar{\sigma}^{m\dot{\beta}\alpha} = -2\eta^{lm}. \tag{1.37}$$

From the completeness of the set $\mathbf{I}$, $\boldsymbol{\sigma}$ for 2×2 matrices we may write for any matrix $\mathbf{A}$

$$\mathbf{A} = \tfrac{1}{2}\mathrm{Tr}\mathbf{A} + \tfrac{1}{2}\mathrm{Tr}(\sigma_i\mathbf{A})\sigma_i$$

or

$$A_{\alpha\dot{\beta}} = \delta^{\sigma}_{\alpha}\delta^{\dot{\rho}}_{\dot{\beta}}A_{\sigma\dot{\rho}} = -\tfrac{1}{2}[\sigma^{0}_{\alpha\dot{\beta}}\bar{\sigma}_{0}^{\dot{\rho}\sigma}A_{\sigma\dot{\rho}} + \sigma^{i}_{\alpha\dot{\beta}}\bar{\sigma}_{i}^{\dot{\rho}\sigma}A_{\sigma\dot{\rho}}]. \tag{1.38}$$

The second completeness relation then follows as

$$\sigma^{l}_{\alpha\dot{\beta}}\bar{\sigma}_{l}^{\dot{\rho}\sigma} = -2\delta^{\sigma}_{\alpha}\delta^{\dot{\rho}}_{\dot{\beta}}. \tag{1.39}$$

A tensorial index 'l' may then be substituted by a pair of spinorial indices by defining

$$v_{\alpha\dot{\alpha}} = \sigma^{l}_{\alpha\dot{\alpha}}v_{l}. \tag{1.40}$$

Then

$$v_{l} = -\tfrac{1}{2}\bar{\sigma}^{l\dot{\alpha}\alpha}v_{\alpha\dot{\alpha}} = -\tfrac{1}{2}\sigma^{l}_{\alpha\dot{\alpha}}v^{\alpha\dot{\alpha}}. \tag{1.41}$$

All the Fierz rearrangements to be discussed in the next section may be derived by using the two completeness relations and the properties of σ^{l}, $\bar{\sigma}^{l}$ matrices.

Some other useful relations are

$$\begin{aligned}\sigma^{l}\bar{\sigma}^{m}\sigma^{n} + \sigma^{n}\bar{\sigma}^{m}\sigma^{l} &= 2(\eta^{ln}\sigma^{m} - \eta^{lm}\sigma^{n} - \eta^{mn}\sigma^{l})\\ \sigma^{l}\bar{\sigma}^{m}\sigma^{n} - \sigma^{n}\bar{\sigma}^{m}\sigma^{l} &= 2\mathrm{i}\varepsilon^{lmnp}\sigma_{p}\end{aligned} \tag{1.42}$$

where $\varepsilon_{0123} = +1$ and ε_{lmnp} is a completely antisymmetric tensor†. The first follows on using (1.19) while the second, say, follows by a direct calculation. The complex conjugate of these relations may be derived using $\bar{\sigma}^{l} = \sigma_2\sigma^{l*}\sigma_2$. We may then derive easily

$$2\mathrm{Tr}\sigma^{mn}\sigma^{kl} = \eta^{nk}\eta^{ml} - \eta^{mk}\eta^{nl} - \mathrm{i}\varepsilon^{mnkl}. \tag{1.43}$$

† $\varepsilon^{pqlm}\varepsilon_{p'q'lm} = -2!(\delta^{p}_{q'}\delta^{q}_{q'} - \delta^{p}_{q'}\delta^{q}_{p'})$, $\varepsilon^{pqlm}\varepsilon_{p'qlm} = -3!\,\delta^{p}_{p'}$ etc.

We also note that $(\sigma^{lm})_\alpha{}^\beta$, $(\bar{\sigma}^{lm})^{\dot{\alpha}}{}_{\dot{\beta}}$ satisfy

$$\sigma^{lm} = \tfrac{1}{2}\mathrm{i}\varepsilon^{lmpq}\sigma_{pq} \qquad \bar{\sigma}^{lm} = -\tfrac{1}{2}\mathrm{i}\varepsilon^{lmpq}\bar{\sigma}_{pq} \tag{1.44}$$

$$(\sigma^l\bar{\sigma}^m)_{\alpha\beta} = \varepsilon_{\beta\rho}(\sigma^l\bar{\sigma}^m)_\alpha{}^\rho = (\sigma^l\bar{\sigma}^m)_{\beta\alpha} \tag{1.45}$$

and the dual ${}^*F_{lm}$ of an antisymmetric tensor F_{lm} is given by

$${}^*F^{lm} = \tfrac{1}{2}\mathrm{i}\varepsilon^{lmpq}F_{pq} \qquad F_{lm} = \tfrac{1}{2}\mathrm{i}\varepsilon_{lmpq}{}^*F^{pq} \tag{1.46}$$

if we require ${}^*({}^*\mathbf{F}) = \mathbf{F}$.

1.2.3 *Spinors in the Weyl representation*

The Dirac spinor takes the form

$$\psi = \begin{pmatrix} \chi_\alpha \\ \bar{\chi}_c^{\dot{\alpha}} \end{pmatrix} \qquad \bar{\psi} = \mathrm{i}(\chi_c^\alpha, \bar{\chi}_{\dot{\alpha}}) \tag{1.47}$$

while

$$\psi_c = \begin{pmatrix} \chi_{c\alpha} \\ \bar{\chi}^{\dot{\alpha}} \end{pmatrix} \qquad \bar{\psi}_c = (\chi^\alpha, \bar{\chi}_{c\dot{\alpha}}). \tag{1.48}$$

Thus under charge conjugation $\chi \leftrightarrow \chi_c$. The Majorana condition reads $\chi_c = \chi$. A 2-spinor χ defines a Majorana spinor

$$\begin{pmatrix} \chi \\ \bar{\chi} \end{pmatrix}$$

while two independent 2-spinors are required to define a Dirac spinor. The chiral projections are

$$\psi_L = \begin{pmatrix} \chi_\alpha \\ 0 \end{pmatrix} \qquad \psi_R = \begin{pmatrix} 0 \\ \bar{\chi}_c^{\dot{\alpha}} \end{pmatrix} \qquad \bar{\psi}_L = \mathrm{i}(0, \bar{\chi}_{\dot{\alpha}})$$

$$\bar{\psi}_R = \mathrm{i}(\chi_c^\alpha, 0). \tag{1.49}$$

For the bilinear forms we find ($\varphi = \begin{pmatrix} \eta \\ \bar{\eta}_c \end{pmatrix}$)

$$\begin{aligned}
\bar{\psi}\varphi &= \mathrm{i}(\chi_c\eta + \bar{\chi}\bar{\eta}_c) \\
\mathrm{i}\bar{\psi}\gamma_5\varphi &= \mathrm{i}(-\chi_c\eta + \bar{\chi}\bar{\eta}_c) \\
\bar{\psi}\gamma^l\varphi &= -(\chi_c\sigma^l\bar{\eta}_c + \bar{\chi}\bar{\sigma}^l\eta) = (-\chi_c\sigma^l\bar{\eta}_c + \eta\sigma^l\bar{\chi}) \qquad (1.50) \\
\mathrm{i}\bar{\psi}\gamma_5\gamma^l\varphi &= (\chi_c\sigma^l\bar{\eta}_c + \eta\sigma^l\bar{\chi}) \\
\bar{\psi}\gamma^l\partial_l\varphi &= -\chi_c\sigma^l\partial_l\bar{\eta}_c + (\partial_l\eta)\sigma^l\bar{\chi}.
\end{aligned}$$

Here we use the following notation and properties

$$\begin{aligned}
\chi\eta &\equiv \chi^\alpha\eta_\alpha = \varepsilon_{\alpha\beta}\chi^\alpha\eta^\beta = -\chi_\alpha\eta^\alpha = \eta^\alpha\chi_\alpha = \eta\chi \\
\bar\chi\bar\eta &\equiv \bar\chi_{\dot\alpha}\bar\eta^{\dot\alpha} = -\bar\chi^{\dot\alpha}\bar\eta_{\dot\alpha} = \bar\eta_{\dot\alpha}\bar\chi^{\dot\alpha} = \bar\eta\bar\chi \\
\chi\sigma^l\bar\eta &= \chi^\alpha\sigma^l{}_{\alpha\dot\beta}\bar\eta^{\dot\beta} = -\bar\eta^{\dot\beta}\bar\sigma^l{}_{\dot\beta\alpha}\chi^\alpha = -\bar\eta_{\dot\beta}\bar\sigma^{l\dot\beta\alpha}\chi_\alpha = -\bar\eta\bar\sigma^l\chi \\
(\sigma^l\bar\eta)^\alpha &= \varepsilon^{\alpha\beta}(\sigma^l\bar\eta)_\beta = \varepsilon^{\alpha\beta}\sigma^l{}_{\beta\dot\rho}\bar\eta^{\dot\rho} = -\bar\eta_{\dot\rho}\bar\sigma^{l\dot\rho\alpha} = -(\bar\eta\bar\sigma^l)^\alpha \text{ etc...}
\end{aligned} \tag{1.51}$$

where fermionic factors as usual are assumed to anticommute.

The most important fact is that the *Fierz rearrangement* may be performed using the simple completeness relations of the 2-spinor formulation. For example,

$$\begin{aligned}
2\xi^\alpha\bar\eta^{\dot\beta} &= 2\xi^\sigma\bar\eta^{\dot\rho}\,\delta^\alpha_\sigma\delta^{\dot\beta}_{\dot\rho} = -\xi^\sigma\bar\eta^{\dot\rho}\sigma^l_{\sigma\dot\rho}\bar\sigma_l^{\dot\beta\alpha} = -(\xi\sigma^l\bar\eta)\bar\sigma_l^{\dot\beta\alpha} \\
2(\sigma^l\bar\eta)_\rho(\xi\chi) &= -2\sigma^l_{\rho\dot\beta}\xi^\alpha\bar\eta^{\dot\beta}\chi_\alpha = 2(\xi\sigma_m\bar\eta)(\sigma^l\bar\sigma^m\chi)_\rho \\
2\xi^\alpha(\chi\sigma^l\bar\eta) &= -2\xi^\alpha\bar\eta^{\dot\beta}\chi^\sigma\sigma^l_{\sigma\dot\beta} = (\xi\sigma_m\bar\eta)(\chi\sigma^l\bar\sigma^m)^\alpha \\
(\xi\sigma^l\bar\chi)(\bar\sigma_l\eta)^{\dot\alpha} &= \xi^\alpha\bar\chi^{\dot\beta}\eta_\beta\sigma^l_{\alpha\dot\beta}\bar\sigma_l^{\dot\alpha\beta} = -2\xi^\alpha\bar\chi^{\dot\beta}\eta_\beta\delta^\beta_\alpha\delta^{\dot\alpha}_{\dot\beta} \\
&= 2(\xi\eta)\bar\chi^{\dot\alpha} \qquad \text{etc.}
\end{aligned} \tag{1.52}$$

The chirality transformation on 2-spinors reads as

$$\chi \to \mathrm{e}^{\mathrm{i}\alpha(x)}\chi \qquad \chi_\mathrm{c} \to \mathrm{e}^{\mathrm{i}\alpha(x)}\chi_\mathrm{c} \tag{1.53}$$

while the parity transformation is

$$\chi'_\alpha = a\sigma^0_{\alpha\dot\beta}\bar\chi^{\dot\beta}_\mathrm{c} \qquad \bar\chi'^{\dot\alpha}_\mathrm{c} = a\bar\sigma^{0\dot\alpha\beta}\chi_\beta. \tag{1.54}$$

The parity transformation thus reverses the sign of the chiral charge.

The *complex conjugation* is defined such that the order of anticommuting fermionic factors is reversed. For example,

$$\begin{aligned}
(\chi\eta)^* &= \eta^*_\alpha\chi^{*\alpha} = \bar\eta_{\dot\alpha}\bar\chi^{\dot\alpha} = \bar\eta\bar\chi = \bar\chi\bar\eta \\
(\chi\sigma^l\bar\eta)^* &= \eta^\beta(\sigma^l_{\alpha\dot\beta})^*\bar\chi^{\dot\alpha} = \eta^\beta\bar\sigma^l_{\dot\alpha\beta}\bar\chi^{\dot\alpha} = \eta^\beta\sigma^l_{\beta\dot\alpha}\bar\chi^{\dot\alpha} = \eta\sigma^l\bar\chi \\
[(\sigma^l\bar\eta)_\alpha]^* &= (\sigma^l_{\alpha\dot\beta})^*\bar\eta^{\dot\beta *} = \bar\sigma^l_{\dot\alpha\beta}\eta^\beta = \eta^\beta\sigma^l_{\beta\dot\alpha} = (\eta\sigma^l)_{\dot\alpha} \\
(\bar\psi\mathbf{M}\varphi)^* &= \varphi^*_\beta(\gamma^0\mathbf{M})^*_{\alpha\beta}\psi_\alpha = \varphi^\dagger(\gamma^0\mathbf{M})^\dagger\psi = \bar\varphi\gamma^0\mathbf{M}^\dagger\gamma^0\psi.
\end{aligned} \tag{1.55}$$

We will also make use of spinorial (left) derivatives like $\partial_\alpha = \partial/\partial\theta^\alpha$, $\bar\partial^{\dot\alpha} = \partial/\partial\bar\theta_{\dot\alpha}$ etc. We define $\partial_\alpha\theta^\beta = \delta^\beta_\alpha$, $\bar\partial_{\dot\alpha}\bar\theta^{\dot\beta} = \delta^{\dot\beta}_{\dot\alpha}$. It follows that $\varepsilon^{\alpha\beta}\partial_\beta = -\partial^\alpha$, $\varepsilon_{\dot\alpha\dot\beta}\bar\partial^{\dot\beta} = -\bar\partial_{\dot\alpha}$ etc. In conformity with the general rule that spinorial factors anticommute the right spinorial derivatives must be defined such that $\partial_\alpha\theta_\beta = -\theta_\beta\overleftarrow{\partial}_\alpha$, $\partial_\alpha\theta_\beta\theta_\gamma = \theta_\beta\theta_\gamma\overleftarrow{\partial}_\alpha$ etc. It follows from the complex conjugation defined above that $\partial^*_\alpha = -\bar\partial_{\dot\alpha}$. For example, $(\partial_\alpha\theta^\beta)^* = \delta_{\dot\alpha}{}^{\dot\beta} = \bar\theta^{\dot\beta}(\overleftarrow{\partial}_\alpha)^* = -(\partial_\alpha)^*\bar\theta^{\dot\beta} = \bar\partial_{\dot\alpha}\bar\theta^{\dot\beta}$, $[\partial_\alpha(\theta^\beta\bar\theta_{\dot\sigma})]^* = \delta_{\dot\alpha}{}^{\dot\beta}\theta_\sigma = -\bar\partial_{\dot\alpha}(\theta_\sigma\bar\theta^{\dot\beta}) = \theta_\sigma\bar\theta^{\dot\beta}(\overleftarrow{\partial}_\alpha)^* = (\partial_\alpha)^*\theta_\sigma\bar\theta^{\dot\beta}$.

1.2.4 Majorana representation

In four-dimensional space–time there exists a Majorana representation $\tilde{\gamma}^l$ where all the gamma matrices become real. It may be obtained by performing a unitary transformation on the Weyl representation

$$\tilde{\gamma}^l = \mathbf{U}\gamma^l\mathbf{U}^\dagger \qquad \tilde{\gamma}^{l\dagger} = \mathbf{U}\gamma^{\dagger l}\mathbf{U}^\dagger \tag{1.56}$$

where

$$\mathbf{U} = \frac{1}{\sqrt{2}}(\mathbf{I} + \mathrm{i}\gamma^2) \qquad \mathbf{U}^\dagger = \frac{1}{\sqrt{2}}(\mathbf{I} - \mathrm{i}\gamma^2).$$

We find

$$\tilde{\gamma}^2 = \gamma^2 \qquad \tilde{\gamma}^0 = \begin{pmatrix} -\mathrm{i}\sigma_2 & \mathbf{0} \\ \mathbf{0} & \mathrm{i}\sigma_2 \end{pmatrix} = -\tilde{\gamma}^{0\mathrm{T}} \qquad \tilde{\gamma}^1 = \begin{pmatrix} \sigma_3 & \mathbf{0} \\ \mathbf{0} & \sigma_3 \end{pmatrix}$$

$$\tilde{\gamma}^3 = -\begin{pmatrix} \sigma_1 & \mathbf{0} \\ \mathbf{0} & \sigma_1 \end{pmatrix}. \tag{1.57}$$

The representation has the property

$$\tilde{\gamma}^0\tilde{\gamma}^l\tilde{\gamma}^0 = \tilde{\gamma}^{l\dagger} = \tilde{\gamma}^{l\mathrm{T}} \qquad \tilde{\gamma}_5 = \mathbf{U}\gamma_5\mathbf{U}^\dagger = \mathrm{i}\gamma^2\gamma_5 = \tilde{\gamma}_5^* = -\tilde{\gamma}_5^{\mathrm{T}}. \tag{1.58}$$

The charge conjugation matrix may clearly be chosen to be $\tilde{\mathbf{C}} = -\tilde{\gamma}^{0\mathrm{T}} = -\tilde{\mathbf{C}}^{\mathrm{T}} = -\mathbf{C}^{*-1}$ so that $\psi_{\mathrm{c}} = \tilde{\mathbf{C}}\tilde{\gamma}^{0\mathrm{T}}\psi^* = \psi^*$, $(\psi_{\mathrm{c}})_{\mathrm{c}} = \psi$ and $\bar{\psi}_{\mathrm{c}} = -\psi^{\mathrm{T}}\tilde{\mathbf{C}}^{-1} = \psi^{\mathrm{T}}\tilde{\gamma}^0$. Some calculations involving a Majorana spinor ψ_{M} in this representation become very simple. For example, $\bar{\psi}_{\mathrm{M}}\tilde{\gamma}^l\psi_{\mathrm{M}} = \psi_{\mathrm{M}}^{\mathrm{T}}(\tilde{\gamma}^0\tilde{\gamma}^l)\psi_{\mathrm{M}}$ vanishes because $\tilde{\gamma}^0\tilde{\gamma}^l \sim \tilde{\mathbf{C}}\tilde{\gamma}^l$ is symmetric and spinor components anticommute. The Majorana representation may not exist for other space–time dimensions.

1.3 Supersymmetry transformations. Super-Poincaré algebra

1.3.1 Supersymmetry generators

The unusual features of spinorial generators of Fermi–Bose supersymmetry may be motivated in the free field theory action of a real scalar field $A(x)$ and a Majorana field $\psi(x)$

$$\int \mathrm{d}^4x\,\mathscr{L} = \int \mathrm{d}^4x\left(-\frac{1}{2}(\partial_l A)(\partial^l A) + \frac{\mathrm{i}}{2}\,[(\partial_l\psi)\sigma^l\bar{\psi} + (\partial_l\bar{\psi})\bar{\sigma}^l\psi]\right). \tag{1.59}$$

We will work in the system of units with $\hbar = c = 1$ so that the dimension of $\mathscr{L}$, written as $[\mathscr{L}]$, measured in units of inverse length is four, $[\mathscr{L}] = 4$. We derive then $[A] = 1$ and $[\psi] = 3/2$. The simplest linear supersymmetry (SS) transformation of A may be defined as

$$\delta A = (\xi\psi + \bar{\xi}\bar{\psi}) \tag{1.60}$$

where ξ is a constant spinor parameter of the transformation. It follows that $[\xi] = [\bar{\xi}] = -1/2$. To write the corresponding SS transformation of ψ we note that the right-hand side must contain ξA with $[\xi A] = 1/2$. We may fill the gap of unity in the dimensions by introducing a space–time derivative to obtain a simple transformation as

$$\delta\psi_\alpha(x) = \mathrm{i}(\sigma^l\bar{\xi})_\alpha\partial_l A \qquad \delta\bar{\psi}^{\dot{\alpha}}(x) = \mathrm{i}(\bar{\sigma}^l\xi)^{\dot{\alpha}}\partial_l A. \tag{1.61}$$

We have chosen the factors appropriately so that the action (1.59) is invariant under these SS transformations. For the commutator of two SS transformations with parameters ξ_1 and ξ_2 we find for the bosonic field

$$\begin{aligned}\delta_2\delta_1 A &= \xi_2\delta_1\psi + \bar{\xi}_2\delta_1\bar{\psi} + \ldots = \mathrm{i}(\xi_2\sigma^l\bar{\xi}_1 - \xi_1\sigma^l\bar{\xi}_2)A + \ldots \\ [\delta_2, \delta_1]A &= 2\mathrm{i}(\xi_2\sigma^l\bar{\xi}_1 - \xi_1\sigma^l\bar{\xi}_2)\partial_l A\end{aligned} \tag{1.62}$$

where the dots indicate the other terms which cancel out in the commutator. The factor 2 in (1.62) derives from our normalisation in definition (1.60).

We define the infinitesimal generators $Q_\alpha, \bar{Q}^{\dot{\alpha}}$ of a SS transformation with† $[Q] = [\bar{Q}] = 1/2$ by

$$\delta(\text{Field}) = -\mathrm{i}(\xi Q + \bar{\xi}\bar{Q})(\text{Field}). \tag{1.63}$$

On the assumption that ξ, $\bar{\xi}$ anticommute with Q, $\bar{Q}$ we may derive

$$\begin{aligned}[\delta_2, \delta_1](\text{Field}) = [&-\xi_2^\alpha\{Q_\alpha, \bar{Q}_{\dot{\beta}}\}\bar{\xi}_1^{\dot{\beta}} + \xi_1^\alpha\{Q_\alpha, \bar{Q}_{\dot{\beta}}\}\bar{\xi}_2^{\dot{\beta}} \\ &+ \xi_2^\alpha\{Q_\alpha, Q_\beta\}\xi_1^\beta + \bar{\xi}_2^{\dot{\alpha}}\{\bar{Q}_{\dot{\alpha}}, \bar{Q}_{\dot{\beta}}\}\bar{\xi}_1^{\dot{\beta}}](\text{Field})\end{aligned} \tag{1.64}$$

Comparing with (1.62) the generators acting on a field $A(x)$ satisfy the following anticommutation relations

$$\{Q_\alpha, \bar{Q}_{\dot{\beta}}\} = -2\mathrm{i}\sigma^l_{\alpha\dot{\beta}}\partial_l \qquad \{Q_\alpha, Q_\beta\} = 0 \qquad \{\bar{Q}_{\dot{\alpha}}, \bar{Q}_{\dot{\beta}}\} = 0. \tag{1.65}$$

The generator P_l of space–time translations may be defined by $\delta_a x^l = \mathrm{i}(a \cdot P)x^l = a^l$ so that $P_l = -\mathrm{i}\partial_l$. The translation invariance of the fields then gives

$$\delta_a A(x) = -a^l\partial_l A = -\mathrm{i}(a \cdot P)A \qquad \delta_a\psi(x) = -a^l\partial_l\psi = -\mathrm{i}(a \cdot P)\psi(x). \tag{1.66}$$

From $[\delta_a, \delta_\xi](\text{Field}) = 0$ we derive, assuming that a^l commutes with $\xi, \bar{\xi}$,

$$[P_l, Q_\alpha] = [P_l, \bar{Q}_{\dot{\alpha}}] = 0 \tag{1.67}$$

† From $[\mathrm{i}(\xi Q + \bar{\xi}\bar{Q})]^* = -\mathrm{i}[(Q_\alpha)^*\bar{\xi}^{\dot{\alpha}} + (\bar{Q}^{\dot{\alpha}})^*\xi_\alpha] = -\mathrm{i}[\bar{\xi}_{\dot{\alpha}}(Q^\alpha)^* + \xi^\alpha(\bar{Q}_{\dot{\alpha}})^*] = \mathrm{i}(\xi Q + \bar{\xi}\bar{Q})$ we find $(\mathrm{i}Q^\alpha)^* = \mathrm{i}\bar{Q}^{\dot{\alpha}}$ and $(\mathrm{i}\bar{Q}_{\dot{\alpha}})^* = \mathrm{i}Q_\alpha$. In the 4-component notation

$$\begin{pmatrix}\mathrm{i}Q_\alpha \\ \mathrm{i}\bar{Q}^{\dot{\alpha}}\end{pmatrix}$$

is a Majorana spinor.

and consequently

$$[P^l P_l, Q] = [P^l P_l, \bar{Q}] = 0. \quad (1.68)$$

The transformation properties of fields under Lorentz rotations $\delta x^l = \lambda^l{}_m x^m = -(i/2)(\lambda_{pq}L^{pq})x^l$ are given by $\delta(\text{Field}) = (i/2)\lambda_{lm}M^{lm}(\text{Field})$ where $M_{lm} = L_{lm}$ for A, $M_{lm} = L_{lm} + i\sigma_{lm}$ for ψ and $M_{lm} = L_{lm} + i\bar{\sigma}_{lm}$ for $\bar{\psi}$. We derive, say, on field A

$$\begin{aligned}[\delta_L, \delta_\xi]A &= \tfrac{1}{2}\lambda_{lm}(\xi\sigma^{lm}\psi + \bar{\xi}\bar{\sigma}^{lm}\bar{\psi}) \\ &= -\tfrac{1}{2}i\lambda_{lm}(\xi\sigma^{lm}Q + \bar{\xi}\bar{\sigma}^{lm}\bar{Q})A. \quad (1.69)\end{aligned}$$

On the other hand

$$[\delta_L, \delta_\xi]A = [\tfrac{i}{2}\lambda_{lm}M^{lm}, -i(\xi Q + \bar{\xi}\bar{Q})]A. \quad (1.70)$$

We derive, assuming that λ_{lm} commutes with $\xi, \bar{\xi}$,

$$\begin{aligned}[M^{lm}, Q_\alpha] &= -i(\sigma^{lm}Q)_\alpha \\ [M^{lm}, \bar{Q}^{\dot{\alpha}}] &= -i(\bar{\sigma}^{lm}\bar{Q})^{\dot{\alpha}}.\end{aligned} \quad (1.71)$$

These relations express the fact that $Q_\alpha, \bar{Q}_{\dot{\alpha}}$ are SL(2, C) spinors. In fact, if we regard $\psi(x)$ as a field operator the expectation value $\langle\Psi|\psi(x)|\Psi\rangle$ is a c-number and we expect it to transform like an unquantised spinor field

$$\langle\Psi'|\psi(x')|\Psi'\rangle = \mathbf{S}_1(\Lambda)\langle\Psi|\psi(x)|\Psi\rangle \quad (1.72)$$

where $|\Psi'\rangle$ is a Lorentz transformed state $|\Psi'\rangle = \mathbf{U}|\Psi\rangle$. We obtain then

$$\psi'(x') = \mathbf{U}^\dagger\psi(x')\mathbf{U} = \mathbf{S}_1(\Lambda)\psi(x) = \mathbf{S}_1(\Lambda)\psi(\Lambda^{-1}x') \quad (1.73)$$

where $\mathbf{U} = \exp(i/2)\lambda_{lm}M^{lm}$. For infinitesimal transformations we derive

$$\begin{aligned}\delta\psi(x) &= -\tfrac{i}{2}\lambda_{lm}[M^{lm}, \psi(x)] = \tfrac{i}{2}\lambda_{lm}[L^{lm} + i\sigma^{lm}]\psi(x) \\ [M^{lm}, \psi(x)] &= -[L^{lm} + i\sigma^{lm}]\psi(x).\end{aligned} \quad (1.74)$$

The following commutators

$$\begin{aligned}[M_{lm}, P_n] &= -i(\eta_{mn}P_l - \eta_{ln}P_m) \\ [P_l, P_m] &= 0\end{aligned} \quad (1.75)$$

expressing the four-vector nature of P_l are similarly obtained.

The generators M_{lm}, P_l, Q_α, $\bar{Q}_{\dot{\alpha}}$ give rise to a closed algebra called super-Poincaré algebra $S\mathscr{P}_4$. It is the minimal extension of Poincaré algebra generated by M_{lm}, P_l containing in addition the spinorial SS generators $Q, \bar{Q}$. The algebra closes under a non-associative modified or graded Lie product defined by

$$M_1 \wedge M_2 = [M_1, M_2\} = M_1M_2 - (-1)^{m_1m_2}M_2M_1 \quad (1.76)$$

where m denotes the total number of spinorial indices contained in M. (See Ne'eman *et al.* 1975, Srivastava 1976b.)

The anticommutator $\{Q_\alpha, \bar{Q}_{\dot\beta}\} = 2\sigma^l_{\alpha\dot\beta}P_l$ leads to an important positive energy theorem for a supersymmetric theory. We derive from (1.37)

$$P^l = -\tfrac{1}{4}\bar{\sigma}^{l\dot\beta\alpha}\{Q_\alpha, \bar{Q}_{\dot\beta}\}. \tag{1.77}$$

For the time component it gives ($\bar{Q}_{\dot\alpha} = -(Q_\alpha)^*$)

$$\begin{aligned} P^0 = -P_0 &= -\tfrac{1}{4}(Q_1\bar{Q}_{\dot 1} + \bar{Q}_{\dot 1}Q_1 + Q_2\bar{Q}_{\dot 2} + \bar{Q}_{\dot 2}Q_2) \\ &= \tfrac{1}{4}[Q_1(Q_1)^* + (Q_1)^*Q_1 + Q_2(Q_2)^* + (Q_2)^*Q_2]. \end{aligned} \tag{1.78}$$

It follows that in an SS theory the energy in any state is positive definite and a supersymmetric ground state carries vanishing energy.

We also note that the algebra remains unaltered under a chiral phase transformation† $Q \to e^{-i\alpha}Q, \bar{Q} \to e^{+i\alpha}\bar{Q}$ discussed earlier in connection with Majorana spinors. We may enlarge the $S\mathscr{P}_4$ algebra by adding to it the corresponding bosonic generator R satisfying

$$[R, Q_\alpha] = -Q_\alpha \qquad [R, \bar{Q}_{\dot\alpha}] = +\bar{Q}_{\dot\alpha} \tag{1.79}$$

while R commutes with P_l, M_{lm}, $[R, P_l] = [R, M_{lm}] = 0$.

We remark, however, that the model (1.59) is not a realisation of $S\mathscr{P}_4$ algebra. We do not obtain the closure (1.62) over the fermionic field even if we use equations of motion (on-shell). In Chapter 2 we shall discuss field theory models which do realise the algebra. We need to improve the model (1.59) by adding to it one more real scalar field.

1.3.2 N-extended supersymmetry

Limiting themselves to flat space–time and Lie group symmetries which excluded fermionic generators Coleman and Mandula (1967) and O'Raifeartaigh (1965a,b) studied earlier the properites of bosonic generators of symmetries in relativistic quantum field theory. They found that any such group of symmetries of the S-matrix is the direct product of the Poincaré group with a compact internal symmetry group. Consequently, no non-trivial unification of space–time symmetries with bosonic internal symmetries could be possible. Moreover, since the internal symmetry generators commute with the mass-squared operator and the generalised spin operator (see §2.1) the component fields or particles contained in a (bosonic) symmetry multiplet carry necessarily the same 'spin' and are mass degenerate. The possibility of a non-trivial unification when anticommuting fermionic generators are also present and the algebra is not Lie, though Lie admissable, was investigated by Haag *et al.* (1975). For Poincaré invariant

† $\theta \to e^{i\alpha}\theta$, $\bar\theta \to e^{-i\alpha}\bar\theta$, $e^{i\alpha R}Qe^{-i\alpha R} = e^{-i\alpha}Q$, $e^{i\alpha R}\bar{Q}e^{-i\alpha R} = e^{i\alpha}\bar{Q}$.

theory in flat space–time they found that the S-matrix theory could admit the following general extended SS algebra with $2N$ SS generators Q^i_α, $\bar{Q}^i_{\dot\alpha}$, $i = 1, 2, \ldots, N$, called $S\mathscr{P}_4{}^N$

$$\{Q^i_\alpha, \bar{Q}^j_{\dot\beta}\} = 2\delta^{ij}\sigma^l_{\alpha\dot\beta}P_l$$
$$\{Q^i_\alpha, Q^j_\beta\} = \varepsilon_{\alpha\beta}Z^{ij} \qquad \{\bar{Q}^i_{\dot\alpha}, \bar{Q}^j_{\dot\beta}\} = \varepsilon_{\dot\alpha\dot\beta}\bar{Z}^{ij} \tag{1.80}$$
$$[Z_{ij}, Z_{kl}] = 0 \qquad [Z_{ij}, \bar{Z}_{kl}] = 0$$

where $Z^{ij} = -Z^{ji} = -V^{ij} + \mathrm{i}U^{ij}$ are $\frac{1}{2}N(N-1)$ complex scalar *central charges*, with $[Z] = 1$ which commute with all the generators in the theory and additional commutation relations for the Poincaré algebra are understood. We remark that we now have modified (or super-Jacobi) identities; there are, however, no further restrictions imposed on Z^{ij}. The 'central charges' may occur only for $N > 1$ and in the theory with a scale parameter. We remark that the theorems above may not apply in curved space–time and in such theories non-compact internal symmetry groups are not excluded and do appear in some recent models.

When $Z_{ij} = 0$ the N-extended SS algebra possesses the invariance (relativistic) under U(N) rotations on the internal indices, $Q^i_\alpha \to U^{ij}Q^j_\alpha$, $\bar{Q}^i_{\dot\alpha} \to U^{ij*}\bar{Q}^j_{\dot\alpha}$. The central charges reduce this symmetry to a subgroup of U(N) as is clear from $\mathbf{Z} \to \mathbf{UZU}^{\mathrm{T}}, \mathbf{Z}^* \to \mathbf{U}^*\mathbf{Z}^*\mathbf{U}^\dagger$ where $\mathbf{Z} = (Z_{ij})$.

We make here a useful digression on the Lie algebra of the $(N \times N)$ matrix (Lie) group. The generators $T^a{}_b$ (where the first index indicates a row, the second a column) may be defined by considering an infinitesimal transformation of a group element $\mathbf{A} = (A^a{}_b)$

$$A^a{}_b = \delta^a{}_b + \lambda^a{}_b + \ldots = \delta^a{}_b - \lambda^c{}_d(T^d{}_c)^a{}_b + \ldots . \tag{1.81}$$

Here the infinitesimal $\lambda^a{}_b$ satisfy certain restrictions corresponding to those on $\mathbf{A}$. For example, if $\det \mathbf{A} = 1$ we find $\lambda^a{}_a = 0$ and $\lambda^\dagger = -\lambda$ or $\lambda^a{}_b{}^* = -\lambda^b{}_a$ if $\mathbf{A}^\dagger\mathbf{A} = 1$. Consequently, there follow certain symmetry relations among the generators e.g., $\mathrm{Tr}\,(T^a{}_b) = (1/N)\delta^a{}_b\,\mathrm{Tr}\,(T^c{}_c)$ in the first case while $(T^a{}_b)^\dagger = T^b{}_a$ in the case of the unitary group. The algebra of $T^a{}_b$ is, however, independent of the symmetry relations and may be derived in view of (1.81) from

$$\lambda^a{}_b\mathbf{B}^{-1}T^b{}_a\mathbf{B} = (\mathbf{B}^{-1}\lambda\mathbf{B})^c{}_dT^d{}_c. \tag{1.82}$$

It follows that

$$\mathbf{B}^{-1}T^a{}_b\mathbf{B} = B^a{}_c(B^{-1})^d{}_bT^c{}_d \tag{1.83}$$

and considering $\mathbf{B}$ infinitesimal we find

$$[T^a{}_b, T^c{}_d] = +\delta^c{}_bT^a{}_d - \delta^a{}_dT^c{}_b. \tag{1.84}$$

Hence this algebra holds for any other representation. We use the same notation $T^a{}_b$ for the corresponding representations of generators. We may also define a tensor operator $R^{c_1c_2\ldots c_q}_{d_1d_1\ldots d_p}$ which transforms as

$$\mathbf{S}(\mathbf{B}^{-1})R^{c_1\ldots c_q}_{d_1\ldots d_p}\mathbf{S}(\mathbf{B}) = B^{c_1}{}_{c_1'}\ldots B^{c_q}{}_{c_q'}(B^{-1})^{d_1'}{}_{d_1}\ldots(B^{-1})^{d_p'}{}_{d_p}R^{c_1'\ldots c_q'}_{d_1'\ldots d_p'} \tag{1.85}$$

where $\mathbf{S}(\mathbf{B})$ is the representative of $\mathbf{B}$ in the realisation concerned. It follows that

$$\begin{aligned}[T^a{}_b, R^{c_1\ldots c_q}_{d_1\ldots d_p}] = &+\delta^{c_1}{}_b R^{ac_2\ldots c_q}_{d_1\ldots d_p} + \delta^{c_2}_b R^{c_1ac_3\ldots c_q}_{d_1\ldots d_p} + \ldots \\ &- \delta^a_{d_1} R^{c_1\ldots c_q}_{bd_2\ldots d_p} - \delta^a_{d_2} R^{c_1\ldots c_q}_{d_1b\ldots d_p} - \ldots .\end{aligned} \tag{1.86}$$

In particular,

$$\begin{aligned}[T^a{}_a, R^{c_1\ldots c_q}_{d_1\ldots d_p}] &= (q-p)R^{c_1\ldots c_q}_{d_1\ldots d_p} \\ [T^a{}_a, R^{c_1\ldots c_q}] &= qR^{c_1\ldots c_q} \\ [T^a{}_a, R_{d_1\ldots d_p}] &= -pR_{d_1\ldots d_p} \\ [T^a{}_a, T^b{}_c] &= 0.\end{aligned} \tag{1.87}$$

The operator $T^a{}_a$ is singled out and by Schur's lemma over an irreducible representation it is a multiple of the identity operator. It is convenient to define an alternative set of generators $\{F^a{}_b, T^a{}_a\}$ by

$$\begin{aligned}&F^a{}_b = T^a{}_b - (1/N)\,\delta^a{}_b T^c{}_c \\ &F^a{}_a = 0 \qquad \mathrm{Tr}\,F^a{}_b = 0 \qquad [T^a{}_a, F^a{}_b] = 0.\end{aligned} \tag{1.88}$$

The tracelessness of $F^a{}_b$ follows from (1.84) and they satisfy the same algebra as the $T^a{}_b$'s

$$[F^a{}_b, F^c{}_d] = \delta^c{}_b F^a{}_d - \delta^a{}_d F^c{}_b \tag{1.89}$$

but

$$[F^a{}_b, R^{\vdots\vdots\vdots}] = [T^a{}_b, R^{\vdots\vdots\vdots}] - (1/N)\,\delta^a{}_b(q-p)R^{\vdots\vdots\vdots}. \tag{1.90}$$

We saw above that the generators of the special group with $\det \mathbf{A} = 1$ are required to satisfy $\mathrm{Tr}\,[T^a{}_b - (1/N)\delta^a{}_b T^c{}_c] = 0$. Since the number of independent generators $F^a{}_b$ is one less, $F^a{}_a = 0$, and they satisfy this restriction as well as (1.89), it is clear that $F^a{}_b$ generate the special subgroup while $T^a{}_a$ generates a one-dimensional Abelian subgroup. For the case of a U(N) group we find $(F^a{}_b)^\dagger = F^b{}_a, (T^a{}_a)^\dagger = T^a{}_a$. Some suitable combinations of these generators may sometimes be more useful. For example, in the case of SU(3) we may define eight Hermitian generators $F_i, i = 1, \ldots, 8$ defined by $F^2{}_1 = (F^1{}_2)^\dagger = -(F_1 + \mathrm{i}F_2)$, $F^3{}_1 = (F^1{}_3)^\dagger = -(F_4 + \mathrm{i}F_5)$, $F^3{}_2 = -(F_6 + \mathrm{i}F_7) = (F^2{}_3)^\dagger$, $\frac{1}{2}(F^1{}_1 - F^2{}_2) = -F_3$, $F^3{}_3 = +(2/\sqrt{3})F_8$. We find that

$$[F_i, F_j] = \mathrm{i}f_{ijk}F_k. \tag{1.91}$$

The structure constants f_{ijk} are real and completely antisymmetric and are easily derived using (1.89). The set of Hermitian generators will be seen to be useful in the discussion of SS gauge theories. We remark that for the unitary group the condition $\mathbf{UU}^{\dagger} = \mathbf{U}^{\dagger}\mathbf{U} = \mathbf{I}$ or equivalently $\mathbf{U}^{-1\mathrm{T}} = \mathbf{U}^{*}$ or $\mathbf{U}^{\dagger -1} = \mathbf{U}$ imply that we have two inequivalent representations in general and, for example, contracting a U(N) vector with its complex conjugate will give a U(N) invariant.

Returning to the discussion of N-extended supersymmetry with vanishing central charges we may identify† $T^{i}{}_{i}$ with the bosonic chiral symmetry generator R so that $[F^{i}{}_{j} + (1/N)\delta^{i}_{j}R]$, where $F^{i}{}_{j}$ generate SU(N), are the generators of U(N). The commutation relations of the bosonic charges $F^{i}{}_{j}$ with Q, $\bar{Q}$ are

$$\begin{aligned} [F^{i}{}_{j}, Q_k] &= -\,[\delta^{i}_{k}Q_j - (1/N)\,\delta^{i}_{j}Q_k] \\ [F^{i}{}_{j}, \bar{Q}^{k}] &= [\delta^{k}_{j}\bar{Q}^{i} - (1/N)\,\delta^{i}_{j}\bar{Q}^{k}] \end{aligned} \tag{1.92}$$

while $F^{i}{}_{j}, R$ commute with the Poincaré generators. It is clear that it is convenient to write the internal symmetry index as an upper index for quantitities like $\bar{Q}^{i}$ transforming in the complex conjugate representation of U(N) so that we may straightforwardly use (1.90).

1.3.3 Local supersymmetry transformations. Supergravitation‡

The SS charges close on to the space–time translations which must be included to have an algebra with the product defined in (1.76). When rigid supersymmetry is raised to a local symmetry so that the parameters become x dependent, $\xi = \xi(x)$, we expect space–time-dependent translations over the 'distance' $\bar{\xi}_1(x)\gamma^{l}\xi_2(x)$ to appear and hence the general coordinate transformations. The gravitation field consequently must also enter in a theory invariant under local supersymmetry. The first successful supergravity theory was formulated by Freedman *et al.* (1976) and Deser and Zumino (1976). The graviton enters the SS theory accompanied by its SS partner, the Rarita–Schwinger Majorana spin-3/2 field called gravitino. The necessity for a gravitino becomes evident when we discuss in the next chapter the spectrum of the component fields or states which must enter together to form a supersymmetry multiplet. The miraculous cancellations of ultraviolet divergences, which plague the conventional gravity theory, by the presence of presumably a short-range gravitino interaction call on us to take the supergravity theory seriously. Such cancellations are the normal features of any theory with Bose–Fermi symmetry. Several attempts at

† Note that we have changed back to notation $i, j, \ldots$ for the internal symmetry index.

‡ See Chapter 11.

supersymmetric grand unified theories which include all the known interactions in nature are also under current research. We refer the reader to the current literature for the actual state of these theories.

1.3.4 Lie-admissibility of the supersymmetry algebra

We have tacitly assumed in the discussion above that in spite of the non-Lie nature of the SS algebra the infinitesimal transformations can be exponentiated to obtain a Lie supergroup. That this is possible derives from the Lie-admissibility of the superalgebra (1.76) (Srivastava 1976a). The Lie-admissible algebra which is an algebraic covering of the Lie algebra was first identified by Albert (1948). It is such a covering which allows a Lie-admissible infinitesimal behaviour while preserving the global structure of a Lie group. Roughly speaking an algebra equipped with an abstract product ab of its elements is Lie-admissible if it gives rise to a Lie algebra under the product $\circ$ defined by $a \circ b = ab - ba$, that is, we require this product to satisfy $a \circ b = -b \circ a$ and the Jacobi identity $(a \circ b) \circ c + (b \circ c) \circ a + (c \circ a) \circ b = 0$ along with the other properties of an algebra. For such algebras a generalisation of Lie's theory in terms of this attached Lie algebra seems possible.

The cross product defined in (1.76), even though the product $M_1 M_2$ is associative, gives rise to a non-associative algebra with the following properties

$$
\begin{aligned}
M_1 \wedge M_2 &= -(-1)^{m_1 m_2} M_2 \wedge M_1 \\
M_1 \wedge (M_2 M_3) &= (M_1 \wedge M_2) M_3 + (-1)^{m_1 m_2} M_2 (M_1 \wedge M_3) \\
(M_1 M_2) \wedge M_3 &= M_1 (M_2 \wedge M_3) + (-1)^{m_2 m_3(} M_1 \wedge M_3) M_2 \\
M_1 \wedge (M_2 \wedge M_3) - (M_1 \wedge M_2) \wedge M_3 &= (-1)^{m_1 m_2} M_2 \wedge (M_1 \wedge M_3)
\end{aligned}
\tag{1.93}
$$

along with the usual relations of an algebra. The last relation may be recast as the modified super-Jacobi identity

$$
\begin{aligned}
M_1 \wedge (M_2 \wedge M_3) &+ (-1)^{m_1(m_2+m_3)} M_2 \wedge (M_3 \wedge M_1) \\
&+ (-1)^{m_3(m_1+m_2)} M_3 \wedge (M_1 \wedge M_2) = 0.
\end{aligned}
\tag{1.94}
$$

If we define now $M_1 \circ M_2 = M_1 \wedge M_2 - M_2 \wedge M_1$ we find

$$
\begin{aligned}
M_1 \circ M_2 = -M_2 \circ M_1 &= c[m_1 m_2] M_1 \wedge M_2 \\
(M_1 \circ M_2) \circ M_3 + (M_2 \circ M_3) \circ M_1 &+ (M_3 \circ M_1) \circ M_2 \\
&= c[m_1 m_2] c[m_3(m_1 + m_2)] (M_1 \wedge M_2) \wedge M_3 \\
&+ c[m_2 m_3] c[m_1(m_2 + m_3)] (M_2 \wedge M_3) \wedge M_1 \\
&+ c[m_3 m_1] c[m_2(m_1 + m_3)] (M_3 \wedge M_1) \wedge M_2
\end{aligned}
\tag{1.95}
$$

where $c[mm'] = 1 + (-1)^{mm'}$ and $m = 1$ for a spinor index while $m = 0$ for a bosonic one. It is easily shown by considering the possible choices for m_i's that the RHS of (1.95) indeed vanishes. The graded Lie algebra of supersymmetry is thus† Lie-admissible and much of the Lie algebra theory may be extended to it with the appropriate modifications; in particular, a connected (super-) Lie group structure persists (Santilli 1978).

Problems

1.1 Demostrate the following:

$$(\theta - \theta')^{\alpha}(\theta - \theta')_{\beta} = \tfrac{1}{2}\delta^{\alpha}_{\beta}(\theta - \theta')^2 \qquad (\bar\theta - \bar\theta')_{\dot\alpha}(\bar\theta - \bar\theta')^{\dot\beta} = \tfrac{1}{2}\delta_{\dot\alpha}{}^{\dot\beta}(\bar\theta - \bar\theta')^2$$

$$\partial^{\alpha}\partial^{\beta} = \tfrac{1}{2}\varepsilon^{\alpha\beta}\partial^2 \qquad \partial_{\alpha}\partial_{\beta} = -\tfrac{1}{2}\varepsilon_{\alpha\beta}\partial^2 \qquad \bar\partial^{\dot\alpha}\bar\partial^{\dot\beta} = -\tfrac{1}{2}\varepsilon^{\dot\alpha\dot\beta}\bar\partial^2$$

$$\bar\partial_{\dot\alpha}\bar\partial_{\dot\beta} = \tfrac{1}{2}\varepsilon_{\dot\alpha\dot\beta}\bar\partial^2 \qquad \theta\cdot\partial = \theta^{\alpha}\partial_{\alpha} = \theta_{\alpha}\partial^{\alpha} \qquad \bar\theta\cdot\bar\partial = \bar\theta_{\dot\alpha}\bar\partial^{\dot\alpha} = \bar\theta^{\dot\alpha}\bar\partial_{\dot\alpha}$$

$$\sigma^l\bar\sigma^m = -\eta^{lm}(\delta_{\alpha}{}^{\beta}) + 2(\sigma^{lm}{}_{\alpha}{}^{\beta})$$

$$(\sigma^{lm}\boldsymbol{\varepsilon}^{\mathrm{T}})_{\alpha\beta} = (\sigma^{lm}\boldsymbol{\varepsilon}^{\mathrm{T}})_{\beta\alpha} \text{ where } \boldsymbol{\varepsilon} = (\varepsilon_{\alpha\beta})$$

$$\chi_{\alpha}\eta_{\beta} = \tfrac{1}{2}\chi\eta\varepsilon_{\alpha\beta} - \tfrac{1}{2}(\sigma^{lm}\boldsymbol{\varepsilon}^{\mathrm{T}})_{\alpha\beta}\chi\sigma_{lm}\eta$$

$$\chi\sigma^{lm}\eta = -\eta\sigma^{lm}\chi$$

$$(\theta\chi)(\theta\eta) = -\tfrac{1}{2}\chi\eta\theta^2 \qquad (\theta\sigma^l\bar\theta)(\theta\sigma^m\bar\theta) = -\tfrac{1}{2}\theta^2\bar\theta^2\eta^{lm}$$

$$[(\sigma_l\bar\theta)_{\alpha}]^* = (\theta\sigma_l)_{\dot\alpha}.$$

1.2 Verify (1.42), (1.43) and (1.44).

1.3 Verify the following relations ($\gamma^{lm} = \tfrac{1}{4}[\gamma^l, \gamma^m], \varepsilon_{0123} = 1$):

$$\gamma^l\gamma^m = \eta^{lm} + 2\gamma^{lm}$$

$$\gamma^l\gamma^m\gamma^n = \eta^{lm}\gamma^n - \eta^{ln}\gamma^m + \eta^{mn}\gamma^l + \varepsilon^{lmnp}\gamma_p\gamma_5$$

$$\{\gamma^l, \gamma^{mn}\} = \varepsilon^{lmnp}\gamma_p\gamma_5$$

$$4\gamma^{lm}\gamma_5 = -\varepsilon^{lmnp}\gamma_n\gamma_p.$$

1.4 Prove the Fierz rearrangement formulae

$$(\bar\chi\mathbf{M}\lambda)(\bar\psi\mathbf{N}\eta) = -\tfrac{1}{4}(\bar\chi O^i\eta)(\bar\psi\mathbf{N}O_i\mathbf{M}\lambda)$$

where $O^i = \{1, \gamma^l, \mathrm{i}\gamma_5, \gamma_5\gamma^l, \sqrt{2}\gamma^{lm}\}$ is the complete basis of Dirac matrices satisfying $\mathrm{Tr}(O^iO_j) = 4\delta^i_j$, e.g., $\mathrm{Tr}\gamma^l\gamma_m = 4\delta^l_m$.
Hints: $M^A{}_BN^C{}_D = (P_B{}^C)^A{}_D = \tfrac{1}{4}a_{iB}{}^C(O^i)^A{}_D$ etc.

† It is Jordan-admissible as well.

1.5 Define the 4-component Majorana spinor supersymmetry generator by

$$\mathbf{Q} = (Q_A) = \begin{pmatrix} \mathrm{i}Q_\alpha \\ \mathrm{i}\bar{Q}^{\dot\alpha} \end{pmatrix} \qquad \bar{\mathbf{Q}} = \mathbf{Q}^{\mathrm{T}*}\gamma^0 = -(Q^\alpha, \bar{Q}_{\dot\alpha}) = (\bar{Q}_A)$$

$$\mathbf{C}\bar{\mathbf{Q}}^{\mathrm{T}} = \mathbf{Q} \qquad \bar{\mathbf{Q}} = -\mathbf{Q}^{\mathrm{T}}\mathbf{C}^{-1}.$$

Show that the SS algebra may be written as

$$\{Q^i_A, Q^j_B\} = 2(\gamma^l C)_{AB} P_l \delta^{ij} \qquad [M_{lm}, Q^i{}_A] = \mathrm{i}(\gamma^{lm})_{AB} Q^i_B$$

$$\{Q^i_A, \bar{Q}^j_B\} = -2(\gamma^l)_{AB} P_l \delta^{ij}$$

$$[Q, R] = -\mathrm{i}\gamma_5 Q.$$

1.6 Show that in the Majorana representation $Q^*_A = Q_A$, $Q_A = (\mathbf{Q}^{\mathrm{T}}\gamma^0)_A = -(\gamma^0\mathbf{Q})_A$ and derive for the extended SS (without central charges)

$$H = P^0 = \tfrac{1}{4} \sum_{A=1}^{4} (Q^1_A)^2 = \ldots = \tfrac{1}{4} \sum_{A=1}^{4} (Q^i{}_A)^2 = \ldots .$$

Show that if one SS is spontaneously broken, e.g. $Q_A{}^1 | 0 > \neq 0$ for at least one A, they are all broken by the same amount.

1.7 Show that $(\psi_{\mathrm{L(R)}})^{\mathrm{c}} = (\psi^{\mathrm{c}})_{\mathrm{R(L)}}$ and for a Majorana spinor θ we find

$$\bar\theta_{\mathrm{L}}\psi = \bar\theta_{\mathrm{L}}\psi_{\mathrm{R}} = (\bar\theta_{\mathrm{R}})_{\mathrm{c}}\psi_{\mathrm{R}} = \mathrm{i}\bar\theta\bar\chi_{\mathrm{c}} \qquad \bar\theta_{\mathrm{R}}\psi = \bar\theta_{\mathrm{R}}\psi_{\mathrm{L}} = (\bar\theta_{\mathrm{L}})_{\mathrm{c}}\psi_{\mathrm{L}} = \mathrm{i}\theta\chi.$$

1.8 Write the kinetic term for a Majorana spinor in terms of its left (right) chiral projection.

1.9 Show that $(T^a{}_b)^c{}_d = -\delta^a_d\delta^c_b$, $(F^a{}_b)^c{}_d = -\delta^a_d\delta^c_b + (1/N)\delta^c_d\delta^a_b$, $G^a{}_b = x^a\partial_b$ and $(x^a\partial_b - (1/N)x\cdot\partial)$ individually realise the algebra (1.84) and $(\mathbf{A} - \mathbf{I})^a{}_b \approx \lambda^a{}_b = -(\lambda^c{}_d T^d{}_c)^a{}_b$ while $\delta x^a = \lambda^a{}_b x^b = (\lambda^c{}_d G^d{}_c)x^a$.

1.10 From the definitions of the Hermitian generators F_i of SU(3) in terms of the generators $F^a{}_b$ and (1.91) derive the numerical values of structure constants f_{ijk}. Find a 3×3 matrix (fundamental) representation of F_i using the previous problem. Write out also the 8×8 matrices $(F_i)_{jk} = -\mathrm{i}f_{ijk}$ of the adjoint representation.

1.11 Derive the algebra of Lorentz generators from that of $T^a{}_b$ in (1.84).

2

REALISATIONS OF SUPERSYMMETRY

2.1 Representation on one-particle states

2.2.1 Massless case

The representations of the Poincaré group are characterised by the Casimir operators $(-P^l P_l)$ and generalised spin operator $W^2 = W^l W_l$ where W^l is the Pauli–Lubanski vector

$$W_l = \tfrac{1}{2}\varepsilon_{lmpq}P^m M^{pq}$$
$$W \cdot P = 0 \qquad [W_l, P_m] = 0 \qquad W_0 = \boldsymbol{J}\cdot\boldsymbol{P}. \tag{2.1}$$

In the case of the super-Poincaré algebra the mass-squared operator is still a Casimir operator and the states (or component fields) in a supermultiplet carry the same mass. However, now W^2 is not a Casimir operator and a supermultiplet contains states (or fields) with different 'spins' characterised by eigenvalues of W^2. We find on using (1.44) and (1.71) that

$$\begin{aligned}[W_l, Q^i_\alpha] &= \tfrac{1}{2}\varepsilon_{lmpq}P^m[M^{pq}, Q^i_\alpha] \\ &= -\tfrac{1}{2}\mathrm{i}\varepsilon_{lmpq}P^m(\sigma^{pq}Q^i)_\alpha = -P^m(\sigma_{lm}Q^i)_\alpha.\end{aligned} \tag{2.2}$$

The generators of the 'little group' which leave the eigenvalue of P_l invariant now contain SS generators Q^i_α, $\bar{Q}^i_\alpha$ along with W_l and other internal symmetry generators.

For the massless case we may always go to the 'standard' frame $\tilde{p}^l = (\omega, 0, 0, \omega)$ where we find $W^l = -\omega(M_{12}, [M_{23} - M_{02}], [M_{31} + M_{01}], M_{12})$, $[W^1, W^2] = 0$, $[J_3, W^1] = \mathrm{i}W^2$, $[J_3, W^2] = -\mathrm{i}W^1$ and $J_3 = M_{12} = W_0/\omega$. The algebra of these generators is $\mathrm{SO(2)} \otimes \mathrm{T}_2$. In order to obtain a discrete spectrum of momentum eigenstates we must set $W_1 = W_2 = 0$. They generate translations and will give rise to a continuous spectrum for 'spin'. Out of W^l we are thus left with J_3 which generates rotations in the 12 plane. The half-integer eigenvalues λ of J_3 $(= \boldsymbol{J}\cdot\boldsymbol{P}/|\boldsymbol{P}| = W_0)$ introduce the discrete quantum number of helicity to label the momentum eigenstates. We have $W_0|\tilde{p},\rangle = \lambda\omega|\tilde{p},\lambda\rangle = -\lambda p_0|\tilde{p},\lambda\rangle$, $W_3|\tilde{p},\lambda\rangle = -\lambda p_3|\tilde{p},\lambda\rangle$, $W_1|\tilde{p},\lambda\rangle = W_2|\tilde{p},\lambda\rangle = 0$ which may hence be written (for massless particles) in an arbitrary frame as $W_l|\tilde{p},\lambda\ldots\rangle = -\lambda P_l|\tilde{p},\lambda\ \ldots\rangle$ or $W_l = -\lambda P_l$. We note that $W^2 = 0$.

The SS algebra in the standard frame reduces to

$$\{Q^i_\alpha, \bar{Q}^j_{\dot\beta}\} = 2\omega\delta^{ij}(\sigma_0+\sigma_3)_{\alpha\dot\beta} = -4\omega\delta^{ij}\begin{pmatrix}0 & 0\\ 0 & 1\end{pmatrix}$$
$$\{Q^i_\alpha, Q^j_\beta\} = \{\bar{Q}^i_{\dot\alpha}, \bar{Q}^j_{\dot\beta}\} = 0. \tag{2.3}$$

We find

$$\{Q^i_1\bar{Q}^j_{\dot 1} + \bar{Q}^j_{\dot 1}Q^i_1\} = 0. \tag{2.4}$$

If we make the 'positive metric assumption' for the Hilbert space of state vectors (that the absolute 'square' of an operator is non-negative, and zero only if the operator is zero itself), we may set $Q^i_1 = \bar{Q}^i_{\dot 1} = 0$. The remaining generators $Q^i_2, Q^i_{\dot 2}$ generate a Clifford algebra for N fermionic degrees of freedom:

$$\{Q^i, \bar{Q}^j\} = \delta^{ij} \qquad \{Q^i, Q^j\} = \{\bar{Q}^i, \bar{Q}^j\} = 0 \tag{2.5}$$

where

$$\bar{Q}^i = \frac{1}{2\sqrt{\omega}}Q^i_2 \qquad Q^i = -\frac{\bar{Q}^i_{\dot 2}}{2\sqrt{\omega}} = \frac{Q^{i\,*}_2}{2\sqrt{\omega}}.$$

From $[M^{12}, Q^i_\alpha] = -\frac{1}{2}(\sigma_3 Q^i)_\alpha$, $[M^{12}, \bar{Q}^{\dot\alpha i}] = -\frac{1}{2}(\sigma_3\bar{Q}^i)^{\dot\alpha}$ we derive

$$J_3 Q^i = Q^i(J_3 - \tfrac{1}{2}) \qquad J_3\bar{Q}^i = \bar{Q}^i(J_3 + \tfrac{1}{2}) \tag{2.6}$$

showing that all Q^i lower the helicity by 1/2 while all $\bar{Q}^i$ raise it.

The physically relevant generators of 'little' algebra are $J_3, Q^i, \bar{Q}^i$ along with internal symmetry generators. Starting from the state $|\lambda_0\rangle$ with the maximum helicity $\lambda_0 > 0$ in a supermultiplet satisfying (we suppress ω)

$$\bar{Q}^i|\lambda_0\rangle = 0 \qquad i = 1, 2, \ldots, N \tag{2.7}$$

we may construct the following lowering chain of states by applying Q^i's on it

$$|\lambda_0\rangle, \left|\lambda_0 - \frac{1}{2}, i_1\right\rangle \ldots, \left|\lambda_0 - \frac{k}{2}, i_1, \ldots, i_k\right\rangle, \left|\lambda_0 - \frac{N}{2}, i_1, \ldots, i_N\right\rangle \tag{2.8}$$

where the multiplicity of the state $|\lambda_0 - k/2, i_1, \ldots, i_k\rangle = Q^{i_1}\ldots Q^{i_k}|\lambda_0\rangle$ is given by

$$\binom{N}{k}$$

since it is completely antisymmetric in the labels $(i_1, \ldots, i_k)$.

Now it is well known that a Lorentz covariant field theory will normally be *CPT* invariant as well. This implies that for every state of a given helicity λ there should be a similar state with the helicity $(-\lambda)$. Indicating by

$|-\lambda_0\rangle$ the state which satisfies

$$Q^i|-\lambda_0\rangle = 0 \tag{2.9}$$

we may obtain the raising chain of states

$$|-\lambda_0\rangle, \left|-\lambda_0+\frac{1}{2}, i_1\right\rangle, \ldots, \left|-\lambda_0+\frac{k}{2}, i_1, \ldots, i_k\right\rangle, \ldots,$$
$$\left|-\lambda_0+\frac{N}{2}, i_1, \ldots, i_N\right\rangle \tag{2.10}$$

where the state $|-\lambda_0+k/2, i_1 \ldots i_k\rangle = \bar{Q}^{i_1} \ldots \bar{Q}^{i_k}|-\lambda_0\rangle$ carries the multiplicity

$$\binom{N}{k}.$$

The consistency requires that $-\lambda_0 + N/2 \leqslant \lambda_0$ or $N \leqslant 4\lambda_0$ or $\lambda_0 \geqslant N/4$. The *CPT* self-conjugate supermultiplet is obtained when $N = 4\lambda_0$ and the two chains coincide. The total multiplicity is 2^N in this case, otherwise it is $2 \cdot 2^N$. For $N \geqslant 9$ we encounter helicities 5/2 and higher which cannot be consistently coupled to spin-2 fields. Moreover, there are more than one helicity-2 states (gravitons). The physical requirements thus allow the maximum of eight independent supersymmetries, $N \leqslant 8$. A supergravity multiplet with one graviton (and N spin-3/2 gravitons) may be constructed for $N \leqslant 8$. The supermultiplets with $N = 8, \lambda_0 = 2$ and $N = 4, \lambda_0 = 1$ are self-conjugates:

$N=8$	$\lambda_0 = 2$ ($N=8$ supergravity):									Total
Helicities:	-2	$-3/2$	-1	$-1/2$	0	1/2	1	3/2	2	
States :	1	8	28	56	70	56	28	8	1	$=256$

$N=4$	$\lambda_0 = 1$ ($N=4$ SS Yang–Mills theory)					Total
Helicities:	-1	$-1/2$	0	1/2	1	
States :	1	4	6	4	1	$=16$

The $N = 7$, $\lambda_0 = 2$ supermultiplet has the same particle content as that of $N = 8$, $\lambda_0 = 2$. The same is true of $N = 3, \lambda_0 = 1$ and $N = 4$, $\lambda_0 = 1$ supermultiplets. Since spin-3/2 does not allow renormalisable couplings we require $N \leqslant 4, \lambda_0 \leqslant 1$ for renormalisable theories.

The smallest supermultiplets are obtained for $N = 1$ with four helicity states, for example,

$\lambda_0 = 1/2$ (Chiral supermultiplet)			
Helicities:	$-1/2$	0	$1/2$
States :	1	1 + 1	1
:	$2(S=0)+(S=1/2)$		

$\lambda_0 = 1$ (Gauge supermultiplet)				
Helicities:	-1	$-1/2$	$1/2$	1
States :	1	1	1	1
:	$(S=1)+(S=1/2)$			

$\lambda_0 = 2$ (Simple supergravity multiplet)				
Helicities:	-2	$-3/2$	$3/2$	2
States :	1	1	1	1
:	$(S=2)+(S=3/2)$			

For $\lambda_0 = 3/2$ we obtain $(S=3/2)+(S=1)$ as the particle content.

2.1.2 *Massive case*

In the absence of 'central charges' the SS algebra in the rest frame, $p^l = (m, \mathbf{0})$, reduces to

$$\{Q^i_\alpha, \bar{Q}^j_{\dot\beta}\} = 2m\,\delta^{ij}\,\sigma_{0\alpha\dot\beta} \qquad \{Q^i_\alpha, Q^j_\beta\} = \{\bar{Q}^i_{\dot\alpha}, \bar{Q}^j_{\dot\beta}\} = 0 \qquad (2.11)$$

which spans the Clifford algebra of $2N$ fermionic degrees of freedom. The generators $W^l = m(0, S)$ now span the SO(3) or angular momentum algebra. The states in a supermultiplet carry the same mass but different spin states. The action of spin-1/2 SS charges on a state $|\,m, s, s_3, \ldots\rangle$ may be obtained by applying the standard angular momentum theory. A chain of states is generated on applying, say $\bar{Q}^i$'s to a Clifford vacuum $|\,m, s_0, s_{03}, \ldots\rangle$ which is annihilated by all the Q^i_α. We refer the reader to the literature (Salam and Strathdee 1975a, Ferrara *et al.* 1981) where the spectrum of massive states has been analysed in detail and even in the presence of central charges which reduce the U(N) symmetry.

2.1.3 *Casimir operators*

For the super-Poincaré group a Casimir operator which generalises W^2 may be easily constructed. We introduce an intrinsic spin vector operator N_l

$$N_l = -\tfrac{1}{2}\bar{\sigma}_l^{\dot{\alpha}\beta}[Q_{\beta i}, \bar{Q}^i_{\dot{\alpha}}] \qquad [N_l, P_m] = 0. \tag{2.12}$$

On using the identity $[AB, C] = A\{B, C\} - \{A, C\}B$ and $\sigma_l\bar{\sigma}_m = -\eta_{lm} + 2\sigma_{lm}$ we find

$$\begin{aligned} [N_l, Q_{\alpha i}] &= 2P_l Q_{\alpha i} + 4P^m(\sigma_{lm}Q)_{\alpha i} \\ [N\cdot P, Q_{\alpha i}] &= 2P^2 Q_{\alpha i}. \end{aligned} \tag{2.13}$$

In view of (2.2) it is then suggested that we define another vector

$$\begin{aligned} &Y_l = W_l + \tfrac{1}{4}N_l \qquad [Y_l, P_m] = 0 \\ &[Y_l, Q_\alpha] = \tfrac{1}{2}P_l Q_\alpha,\ [Y_l, \bar{Q}_{\dot{\alpha}}] = -\tfrac{1}{2}P_l\bar{Q}_{\dot{\alpha}} \\ &Y\cdot P = \tfrac{1}{4}N\cdot P. \end{aligned} \tag{2.14}$$

Clearly

$$[Y_l P_m - Y_m P_l, Q] = [Y_l P_m - Y_m P_l, \bar{Q}] = 0 \tag{2.15}$$

and the square of $(Y_l P_m - Y_m P_l)$

$$Y^2P^2 - (Y\cdot P)^2 \tag{2.16}$$

commutes with all the generators of the algebra $S\mathscr{P}_4{}^N$. The Casimir operator (2.16) defines a quantum number, the superspin $\varkappa$,

$$Y^2 - \frac{1}{P^2}(Y\cdot P)^2 = m^2\varkappa(\varkappa + 1) \tag{2.17}$$

when mass is non-vanishing.

For the massless case it follows from the SS algebra and the 'positive metric assumption' that $P_l\bar{\sigma}^{l\dot{\alpha}\alpha}Q_{\alpha i} = P_l\sigma^{l\alpha\dot{\alpha}}\bar{Q}^i_{\dot{\alpha}} = 0$ which reduces to $Q^i_1 = \bar{Q}^i_{\dot{1}} = 0$ in the standard frame considered earlier. We find from (2.2) and (2.13)

$$\begin{aligned} [W_l, Q_\alpha] &= -\tfrac{1}{2}P_l Q_\alpha \\ [N_l, Q_\alpha] &= 4P_l Q_\alpha \\ [Y_l, Q_\alpha] &= \tfrac{1}{2}P_l Q_\alpha. \end{aligned} \tag{2.18}$$

Going to the standard frame we find $N_1 = N_2 = 0$, and $N^l N_l = 0$ and W_l already calculated. We may hence define for the sector where W_1, W_2 annihilate the states a (discrete) superhelicity quantum number, say, by the relation

$$Y_l + \tfrac{1}{2}RP_l = -\lambda P_l \tag{2.19}$$

where we remind ourselves that $[RP_l, Q_\alpha] = -P_l Q_\alpha$.

We also find that $[R + (1/2P^2)(N \cdot P)]$ commutes with $Q, \bar{Q}$ and defines superchiral charge in the massive case. For the massless case it follows from (2.18) that $[(RP_l + \frac{1}{4}N_l), Q] = 0$ and we may write $RP_l + \frac{1}{4}N_l = g_5 P_l$. The SS invariant generalisation of SU(N) generators which involve the commutator $[Q_{\beta i}, \bar{Q}^j_{\dot\alpha}]$ as well may also be constructed (see problems 2.1, 2.2).

2.2 Realisation on component fields

2.2.1 Wess–Zumino action† (on-shell)

The discussion in §2.1 on the helicity content of the $N=1, \lambda_0 = 1/2$ chiral supermultiplet suggests that the simplest model used in §1.3 to motivate the SS algebra should be improved by adding to it another real scalar field. The SS action thus obtained should be able to realise the SS algebra on the fermionic field as well. We may as well work with a complex scalar field $A(x)$ plus the Majorana spinor field $\psi(x)$ and write the sum of kinetic terms as

$$\mathscr{L} = -(\partial_l \bar{A})(\partial^l A) + \frac{\mathrm{i}}{2}[(\partial_l \psi)\sigma^l \bar{\psi} + (\partial_l \bar{\psi})\bar{\sigma}^l \psi]. \tag{2.20}$$

The following SS transformation may be shown to leave the action invariant

$$\begin{aligned} \delta A &= -\sqrt{2}\,\xi\psi \\ \delta\psi &= -\mathrm{i}\sqrt{2}(\sigma_l \bar{\xi})\partial^l A \\ \delta\bar{\psi} &= -\mathrm{i}\sqrt{2}(\bar{\sigma}_l \xi)\partial^l \bar{A}. \end{aligned} \tag{2.21}$$

The commutator of two SS transformations on a bosonic field gives

$$\begin{aligned} [\delta_2, \delta_1] A &= -\sqrt{2}(\xi_2 \delta_1 \psi - \xi_1 \delta_2 \psi) \\ &= 2\mathrm{i}(\xi_2 \sigma_l \bar{\xi}_1 - \xi_1 \sigma_l \bar{\xi}_2)\partial_l A. \end{aligned} \tag{2.22}$$

The reality of the expression multiplying $(\partial_l A)$ shows that each of the two real components of A realises the SS algebra individually. On the fermionic field we obtain

$$\begin{aligned} [\delta_2, \delta_1]\psi &= -\mathrm{i}\sqrt{2}\sigma^l \partial_l(\bar{\xi}_2 \delta_1 A - \bar{\xi}_1 \delta_2 A) \\ &= 2\mathrm{i}[(\sigma_l \bar{\xi}_2)(\xi_1 \partial^l \psi) - (\sigma_l \bar{\xi}_1)(\xi_2 \partial^l \psi)]. \end{aligned} \tag{2.23}$$

† Wess and Zumino (1974a).

On using the Fierz rearrangement theorems in (1.52) we get

$$[\delta_2, \delta_1]\psi = 2\mathrm{i}(\xi_2\sigma^l\bar{\xi}_1 - \xi_1\sigma^l\bar{\xi}_2)[\partial_l\psi + \sigma_l\bar{\sigma}_m\partial^m\psi]$$
$$\overset{\circ}{=} 2\mathrm{i}(\xi_2\sigma^l\bar{\xi}_1 - \xi_1\sigma^l\bar{\xi}_2)\partial_l\psi. \qquad (2.24)$$

The last equality is obtained if we use the equation of motion $\mathrm{i}\bar{\sigma}^l\partial_l\psi = 0$. We thus obtain an on-shell realisation of the SS algebra in the field theory given by the action in (2.20).

2.2.2 Auxiliary fields. Supermultiplets of currents and anomalies

It is clearly desirable to have supermultiplets of Bose and Fermi component fields, with SS transformations defined on them, such that the SS algebra is realised individually on each of the components independent of any equations of motion. This will allow us to build a tensor calculus for the supermultiplets and facilitate the building of a quantised theory where the fields must be taken away from their classical paths (off-shell). For simple or $N=1$ supersymmetry this may usually be achieved if we add *auxiliary fields* in sufficient number so as to ensure that the number of bosonic degrees of freedom matches that of the fermionic degrees of freedom. The auxiliary fields also play an important role in connection with the spontaneous breakdown of supersymmetry when some of them must acquire a non-zero vacuum expectation value (see Chapter 6).

For the case of an $N=1$ chiral supermultiplet with physical or dynamical fields A and ψ we need one complex auxiliary scalar field F to match the four fermionic degrees of freedom of the Majorana spinor field ψ. The SS algebra is realised on each of the components of the chiral supermultiplet (A, ψ, F) if we define their SS transformations as follows†:

$$\begin{aligned} \delta A &= -\sqrt{2}\,\xi\psi \\ \delta\psi_\alpha &= -\sqrt{2}[\xi_\alpha F + \mathrm{i}(\partial_l A)(\sigma^l\bar{\xi})_\alpha] \\ \delta F &= \mathrm{i}\sqrt{2}\,\partial_l\psi\sigma^l\bar{\xi}. \end{aligned} \qquad (2.25)$$

They may be derived starting from the definition of δA, using dimensional considerations such as we used in §1.3 and imposing the closure of the commutator on each component field. We will derive them in a straightforward fashion in Chapter 4 using superfields. We note that the canonical dimension of the auxiliary is two, $[F] = 2$. We verify easily

$$\begin{aligned} [\delta_2, \delta_1]\psi &= -\sqrt{2}[\xi_2\delta_1 F + \mathrm{i}\,\sigma^l\bar{\xi}_2\partial_l\delta_1 A] - (1 \leftrightarrow 2) \\ &= 2\mathrm{i}[\xi_2\sigma^l\bar{\xi}_1\partial_l\psi + \{\xi_2\sigma_m\bar{\xi}_1 + \xi_1\sigma_m\bar{\xi}_2\}\sigma^l\bar{\sigma}^m\partial_l\psi - (1 \leftrightarrow 2) \\ &= 2\mathrm{i}(\xi_2\sigma^l\bar{\xi}_1 - \xi_1\sigma^l\bar{\xi}_2)\partial_l\psi \end{aligned} \qquad (2.26)$$

† We write ξ to the left and $\bar{\xi}$ to the right of the fields.

and

$$[\delta_2, \delta_1] F = \mathrm{i}\sqrt{2}\, \partial_l \delta_1 \psi \sigma^l \bar{\xi}_2 - (1 \leftrightarrow 2)$$
$$= 2\mathrm{i}(\xi_2 \sigma^l \bar{\xi}_1 - \xi_1 \sigma^l \bar{\xi}_2) F. \tag{2.27}$$

The SS algebra implies that all the physical fields of a supermultiplet must carry the same mass. The field ψ, for example, may describe the quark matter field while its supersymmetric partner A, called squark, a scalar quark field. The SS must somehow be broken to make squark heavy while keeping quark light.

The field theory Lagrangian for the Wess–Zumino scalar multiplet (A, ψ, F), which is renormalisable, contains a mass term and gives supersymmetry invariant action, may be written as†

$$\begin{aligned} \mathscr{L} &= \mathscr{L}_{\mathrm{KE}} + \mathscr{L}_{\mathrm{m}} + \mathscr{L}_{\mathrm{int}} \\ \mathscr{L}_{\mathrm{KE}} &= -(\partial_l \bar{A})(\partial^l A) + \frac{\mathrm{i}}{2}[(\partial_l \psi)\sigma^l \bar{\psi} + (\partial^l \bar{\psi})\bar{\sigma}^l \psi] + \bar{F}F \\ \mathscr{L}_{\mathrm{m}} &= m(AF - \tfrac{1}{2}\psi\psi) + \mathrm{CC} \\ \mathscr{L}_{\mathrm{int}} &= g(A^2 F - A\psi\psi) + \mathrm{CC}. \end{aligned} \tag{2.28}$$

We derive under SS transformations (2.25)

$$\delta \quad = \partial_l K^l \tag{2.29}$$

where

$$K^l = \sqrt{2}(\partial_l \bar{A})\xi\psi - \frac{\mathrm{i}}{\sqrt{2}}\, \xi\sigma_l \bar{\psi} F + \frac{1}{\sqrt{2}}\, \xi\sigma_m \bar{\sigma}_l \partial^m \bar{A}$$
$$- \mathrm{i}\sqrt{2}(\xi\sigma^l \bar{\psi})(m\bar{A} + g\bar{A}^2) + \mathrm{CC}. \tag{2.30}$$

That $\delta\mathscr{L}$ is always a total divergence and never zero is a general result for supersymmetric theories and derives from the commutator relations like (2.26) on each of the component fields.

The equations of motion follow as

$$\begin{aligned} \bar{F} &= -mA - gA^2 \\ \mathrm{i}\sigma^l \partial_l \bar{\psi} &= -m\psi - 2gA\psi \qquad \mathrm{i}\bar{\sigma}^l \partial_l \psi = -(m + 2g\bar{A})\bar{\psi} \\ \Box \bar{A} &= -mF - g(2AF - \psi\psi) \end{aligned} \tag{2.31}$$

plus their complex conjugates. For the auxiliary field with $[F] = 2$ they are algebraic and may be used to eliminate F. We obtain for the interaction and quadratic mass terms

$$\bar{F}F - \tfrac{1}{2}m\psi\psi - mA(m\bar{A} + g\bar{A}^2) - gA\psi\psi - gA^2(m\bar{A} + g\bar{A}^2) + \mathrm{CC}$$
$$= -\tfrac{1}{2}m(\psi\psi + \bar{\psi}\bar{\psi}) - |mA + gA^2|^2 - g(\psi\psi A + \bar{\psi}\bar{\psi}\bar{A}). \tag{2.32}$$

† We will derive it using superfields in Chapter 4.

The action (2.28) is thus a supersymmetric extension of scalar φ^3, φ^4 theory. The invariance under SS imposes a precise relationship among the various coupling constants and is responsible for the high degree of renormalisability of the theory. Moreover, this relationship is found to be stable under renormalisation. There occur 'miraculous cancellations' of quadratic divergences arising from bosonic and fermionic loops contributing with opposite signs. For the Wess–Zumino action only a common wavefunction renormalisation is required for the component fields to absorb the logarithmically divergent infinity and no separate mass and coupling constant renormalisation are needed (non-renormalisation theorem).

The conserved Noether spinor supercurrent, a vector Majorana spinor, is given by† ($\delta\mathscr{L} = \partial_l K^l$)

$$(\xi J^l + \bar{\xi}\bar{J}^l) = \delta\psi^\alpha \frac{\delta\mathscr{L}}{\delta(\partial_l\psi^\alpha)} + \delta\bar{\psi}_{\dot{\alpha}} \frac{\delta\mathscr{L}}{\delta(\partial_l\bar{\psi}_{\dot{\alpha}})} + \delta A \frac{\delta\mathscr{L}}{\delta(\partial_l A)} + \delta\bar{A} \frac{\delta\mathscr{L}}{\delta(\partial_l\bar{A})} - K^l. \tag{2.33}$$

We find

$$\begin{aligned} \frac{1}{\sqrt{2}} J^l &= -\sigma^m\bar{\sigma}^l\psi\partial_m\bar{A} + \mathrm{i}(m\bar{A} + g\bar{A}^2)\sigma^l\bar{\psi} \\ \frac{1}{\sqrt{2}} \bar{J}^l &= -\bar{\sigma}^m\sigma^l\bar{\psi}\partial_m A + \mathrm{i}(mA + gA^2)\bar{\sigma}^l\psi. \end{aligned} \tag{2.34}$$

The currents carry canonical dimension 7/2 and their on-shell conservation, $\partial_l J^l \stackrel{\circ}{=} 0$, may be verified. The current may be 'improved' by adding to J^l_α terms of the form $\partial_m a^{lm}_\alpha$ with $a^{lm} = -a^{ml}$. Such a term does not affect either the conservation law or the charge related to the current J^l. However, such an improvement does affect, for example, the trace of the energy–momentum tensor of a scalar field‡ which becomes nicer. The spinorial charges $Q = \int \mathrm{d}^3x\, J^0$, $\bar{Q} = \int \mathrm{d}^3x\, \bar{J}^0$ satisfy the SS algebra as a consequence of the canonical commutation relations of field operators and generate their transformations under supersymmetry.

We consider now the realisation of the R-transformation (1.79) on the supermultiplet (A, ψ, F). Defining the corresponding generator R by

$$\delta_R(\text{Field}) = -\mathrm{i}\alpha R(\text{Field}) \qquad \alpha \text{ real} \tag{2.35}$$

we find from (1.63) and (1.79) that

$$[\delta_R, \delta_S](\text{Field}) = -\alpha(-\xi Q + \bar{\xi}\bar{Q})(\text{Field}). \tag{2.36}$$

From the discussion on the chiral (or γ_5) transformations in §1.2 which can be consistently defined for Majorana spinors and the arguments which led

† See problem 2.6.

‡ In fact, we have a family of stress tensors $\theta_{lm} = [(\partial_l\varphi) - \frac{1}{2}\eta_{lm}(\partial\varphi)^2] + \xi(\Box\eta_{lm} - \partial_l\partial_m)\varphi^2$; $\partial_l\theta^{lm} = 0$. For $\xi = \frac{1}{6}$ we get $\theta^l{}_l \stackrel{\circ}{=} 0$.

to (1.79) it is suggested that we define for ψ

$$\delta_R \psi = \mathrm{i}\alpha q \psi \tag{2.37}$$

where q is the corresponding chiral charge. The chiral or R-charge for the other component fields may be related to the chiral charge of ψ as follows. We obtain from (2.36)

$$\delta_R \delta_S A - \delta_S \delta_R A = \alpha \xi Q A = \mathrm{i}\alpha \delta_S A \tag{2.38}$$

where we use (1.63) and (2.25) so that $\bar{\xi}\bar{Q}A = 0$ and $\delta_S A = -\mathrm{i}\xi Q A$. If q' indicates the chiral charge of $A, \delta_R A = \mathrm{i}\alpha q' A$, we obtain from (2.25), (2.37) and (2.38) that

$$\mathrm{i}\alpha q' \delta_S A - \mathrm{i}\alpha q \delta_S A = \mathrm{i}\alpha \delta_S A \tag{2.39}$$

which leads to $q' = (q+1)$. Analogously we find the chiral charge q'' of F to be $q'' = (q-1)$. The R-transformation rotates scalars into pseudoscalars and the bosonic and fermionic fields transform differently

$$\begin{aligned}(A, \psi, F) \;&\overset{R}{\rightarrow}\; \mathrm{e}^{\mathrm{i}\alpha(q+1)}(A, \mathrm{e}^{-\mathrm{i}\alpha}\psi, \mathrm{e}^{-\mathrm{i}2\alpha}F)\\ &= (A', \psi', F').\end{aligned} \tag{2.40}$$

The R-transformed supermultiplet is again a chiral supermultiplet which transforms under SS transformations according to (2.25) with the chirally rotated parameters $\xi' = \mathrm{e}^{\mathrm{i}\alpha}\xi, \bar{\xi}' = \mathrm{e}^{-\mathrm{i}\alpha}\bar{\xi}$.

We may compute the R or chiral Noether current for the Wess–Zumino Lagrangian (2.28). It is given by

$$\begin{aligned}\alpha J_5{}^l &= \delta\psi \frac{\delta \mathscr{L}}{\delta(\partial_l \psi)} + \delta A \frac{\delta \mathscr{L}}{\delta(\partial_l A)} + \mathrm{CC}\\ &= -\alpha[\tfrac{1}{2}q(\psi\sigma^l\bar{\psi} - \bar{\psi}\bar{\sigma}^l\psi) + \mathrm{i}(q+1)(A\partial^l\bar{A} - \bar{A}\partial^l A)]\\ &= -\alpha[\tfrac{1}{2}q\bar{\Psi}\mathrm{i}\gamma_5\gamma^l\Psi + \mathrm{i}(q+1)(A\partial^l\bar{A} - \bar{A}\partial^l A].\end{aligned} \tag{2.41}$$

The kinetic term is R-invariant and we find for the total divergence

$$\partial_l J_5{}^l = \delta_R \mathscr{L} = \delta_R \mathscr{L}_\mathrm{m} + \delta_R \mathscr{L}_\mathrm{int} \tag{2.42}$$

$$\begin{aligned}\delta_R \mathscr{L}_\mathrm{m} &= m \cdot 2\mathrm{i}q(AF - \bar{A}\bar{F} - \tfrac{1}{2}\psi\psi + \tfrac{1}{2}\bar{\psi}\bar{\psi})\\ \delta_R \mathscr{L}_\mathrm{int} &= g \cdot \mathrm{i}(3q+1)[A^2 F - \bar{A}^2\bar{F} - A\psi\psi + \bar{A}\bar{\psi}\bar{\psi}].\end{aligned} \tag{2.43}$$

The requirement of R-invariance would not allow the terms $\mathscr{L}_\mathrm{m}$ and $\mathscr{L}_\mathrm{int}$ to be present simultaneously. For $q = 0$ the latter must be absent while for $q = -1/3$ the mass term should be excluded. We note that for the chiral charge $(q+1) = 2/3$ the divergence of the axial R-current (2.42) vanishes when $m \rightarrow 0$. We have in fact

$$\partial_l J_5{}^l \overset{\circ}{=} -\frac{2\mathrm{i}}{3} m(AF - \bar{A}\bar{F} - \tfrac{1}{2}\psi\psi + \tfrac{1}{2}\bar{\psi}\bar{\psi}). \tag{2.44}$$

An 'improved' version of the conserved supersymmetry current such that its 'γ-trace', $\gamma^l J_l$, vanishes when $m \to 0$ may easily be constructed. We find from (2.34)

$$\frac{1}{\sqrt{2}}\sigma_l \bar{J}^l = -2\sigma^m \bar{\psi}\partial_m A - 4\mathrm{i}(mA + gA^2)\psi \tag{2.45}$$

where we use $\sigma_l \bar{\sigma}^l = -4\mathbf{l}$ and $\sigma_l \bar{\sigma}^m \sigma^l = 2\sigma^m$ following respectively from (1.39) and (1.42). On using the equations of motion (2.31) and $\sigma_l \bar{\sigma}^{lm} = -\frac{3}{2}\sigma^m$ we may rewrite it as follows:

$$\begin{aligned}\frac{1}{\sqrt{2}}\sigma_l \bar{J}^l &= -2\mathrm{i}mA\psi - 2\partial_m(\sigma^m \bar{\psi} A) \\ &= -2\mathrm{i}mA\psi + \tfrac{4}{3}\partial_m(\sigma_l \bar{\sigma}^{lm}\bar{\psi}A).\end{aligned} \tag{2.46}$$

The 'improved' version is then clearly given by

$$\begin{aligned}&\frac{1}{\sqrt{2}}\hat{\bar{J}}^l = \frac{1}{\sqrt{2}}\bar{J}^l - \frac{4}{3}\bar{\sigma}^{ln}\partial_n(\bar{\psi}A) \\ &\partial_l \hat{\bar{J}}^l \stackrel{\circ}{=} 0 \qquad \frac{1}{\sqrt{2}}\sigma_l \hat{\bar{J}}^l = -2\mathrm{i}mA\psi.\end{aligned} \tag{2.47}$$

In four-component notation with $A = -(1/\sqrt{2})(\tilde{A} - \mathrm{i}\tilde{B})$, $\tilde{A}$ and $\tilde{B}$ real, it reads

$$\begin{aligned}&J_l = \{\gamma^n \partial_n(\tilde{A} - \tilde{B}\gamma_5) + m(\tilde{A} + \tilde{B}\gamma_5) + \frac{g}{\sqrt{2}}(\tilde{A} + \tilde{B}\gamma_5)^2\}\gamma_l \psi \\ &\hat{J}_l = J_l + \tfrac{1}{3}\partial^n([\gamma_l, \gamma_n](\tilde{A} + \tilde{B}\gamma_5)\psi) \\ &\gamma^l \hat{J}_l = 2m(\tilde{A} - \tilde{B}\gamma_5)\psi \qquad \partial^l \hat{J}_l = 0.\end{aligned} \tag{2.48}$$

The supersymmetric variation of the supercurrent is given by

$$\delta J^l_\alpha = -\mathrm{i}(\xi Q + \bar{\xi}\bar{Q})J_\alpha{}^l = [J^l_\alpha, -\mathrm{i}(\xi Q + \bar{\xi}\bar{Q})] \tag{2.49}$$

where the quantised field operators are understood when the commutator version is used. The SS charges are realised as a space integral of the supercharge density J^0_α, viz $Q_\alpha = \int \mathrm{d}^3 x J^0_\alpha$, if the supercurrent is conserved and the surface integral may be ignored. On integrating (2.49) over $\mathrm{d}^3 x$ we obtain

$$\int \mathrm{d}^3 x \delta J^0_\alpha = [Q_\alpha, -\mathrm{i}\bar{\xi}\bar{Q}] = -2\mathrm{i}(\sigma^l \bar{\xi})_\alpha P_l. \tag{2.50}$$

The four-momentum is itself the space integral over the components of the (symmetric) energy–momentum or stress tensor, $\theta_{mn} = \theta_{nm}$, $\partial^m \theta_{nm} = 0$

$$P^n = \int \mathrm{d}^3 x \theta^{0n} \qquad P_n = -\int \mathrm{d}^3 x \theta_{0n}. \tag{2.51}$$

The SS transformation law of the vector–spinor current J^n_α thus leads to the stress tensor and it is suggested that J^n_α and θ_{mn} belong to the same supersymmetry multiplet. According to the Einstein equations the stress tensor is the source of the gravitational field. In a supergravity theory† we expect then that the supercurrent density will be the source of some other field which is related under SS transformations to the gravitational field. This field is called the spin-3/2 'gravitino' field (see also §2.1.1). We now turn our attention to an 'improved' version of the stress tensor in order to elucidate on a supermultiplet of currents.

The space–time translation invariance leads to a conserved energy-momentum tensor‡

$$\tau_{lm} = (\partial_m A)(\partial_l \bar{A}) + (\partial_m \bar{A})(\partial_l A) - \frac{i}{2}[(\partial_m \psi)\sigma_l \bar{\psi} + (\partial_m \bar{\psi})\bar{\sigma}_l \psi] + \eta_{lm}\mathscr{L}_{WZ}. \tag{2.52}$$

Following the well known procedure it may be symmetrised (Belinfante 1940) and further improved (Callan *et al.* 1970, Srivastava 1973) such that the trace $\theta^m{}_m$ becomes 'nice' so that it also vanishes when $m \to 0$. We find

$$\begin{aligned} \theta_{lm} = \theta_{ml} = (\partial_m A)(\partial_l \bar{A}) + (\partial_m \bar{A})(\partial_l A) - \tfrac{1}{3}(\partial_l \partial_m - \eta_{lm}\Box)A\bar{A} \\ + \frac{i}{4}[\bar{\psi}\bar{\sigma}_l \partial_m \psi + \psi \sigma_l \partial_m \bar{\psi} + (l \leftrightarrow m)] + \eta_{lm}\mathscr{L} \\ \theta^l{}_l \overset{\circ}{=} m[FA + \bar{F}\bar{A} - \tfrac{1}{2}\psi\psi - \tfrac{1}{2}\bar{\psi}\bar{\psi}] \\ \partial^m \theta_{lm} \overset{\circ}{=} 0. \end{aligned} \tag{2.53}$$

Combining (2.44) and (2.53) we observe

$$\tfrac{2}{3}\theta^l{}_l + i\partial_l J_5{}^l = \tfrac{2}{3}m(2FA - \psi\psi) \tag{2.54}$$

where $\theta^m{}_m$ and $\partial_n J_5{}^n$ are real quantities. Now it is an easy exercise to show that $(A^2, 2A\psi_\alpha, 2AF - \psi\psi)$ is also a chiral supermultiplet transforming according to (2.25) and it follows from (2.44) and (2.54) that

$$(\tfrac{2}{3}mA^2, i(\sqrt{2}/3)(\sigma^l \bar{J}_l)_\alpha, \tfrac{2}{3}\theta^l{}_l + i\partial_n J_5{}^n) \tag{2.55}$$

constitutes an (irreducible) chiral 'multiplet of anomalies' (Ferrara and Zumino 1975, Howe *et al.* 1981, Sohnius and West 1981, 1983).
This multiplet vanishes when $m \to 0$ at the classical level. Quantum corrections are, however, expected to give anomalous contributions even for the massless case. One may also verify that mA^2, $J_5{}^n$, $J_\alpha{}^n$, $\bar{J}_n{}^{\dot{\alpha}}$, θ_{mn} themselves are components of a reducible supermultiplet of currents; we may define SS transformations such that the set is closed under them while $\partial_m J_\alpha{}^m = \partial_m J_5{}^m = 0$, $\theta_{mn} = \theta_{nm}$ and $\partial^n \theta_{mn} = 0$ are preserved. For $m = 0$ when $\gamma_m J^m = \theta^m{}_m = 0$ the smaller set $J_5{}^m$, $J_\alpha{}^m$, $\bar{J}_{\dot{\alpha}}{}^m$, θ_{mn} describes an irreducible 'conformal current supermultiplet'.

† See Chapter 11.
‡ In our metric $\tau_{lm} = -(\partial_m \varphi)\delta\mathscr{L}/\delta(\partial^l \varphi) + \eta_{lm}\mathscr{L}$ where φ is a field multiplet.

2.2.3 *Conformal supersymmetry*

The massless Wess–Zumino action possesses more symmetries besides the super-Poincaré symmetry; it is invariant under the graded extension of the conformal group—the super-conformal group. We recall that the (improved) conserved symmetric stress tensor allows us to introduce the dilatation and special conformal currents as moments of θ_{mn}

$$d_n = x^m \theta_{nm} \qquad \partial^n d_n \stackrel{\circ}{=} \theta^n{}_n \qquad D = \int \mathrm{d}^3 x\, x^n \theta_{0n}$$
$$\varkappa_{lm} = (2x_l x^n \theta_{nm} - x^2 \theta_{lm}) \qquad \partial^m \varkappa_{lm} \stackrel{\circ}{=} 2x_l \theta^m{}_m \qquad K_l = \int \mathrm{d}^3 x \varkappa_{l0} \tag{2.56}$$

apart from $M_{lmn} = (x_l \theta_{mn} - x_m \theta_{ln})$, $\partial^n M_{lmn} \stackrel{\circ}{=} 0$, $M_{mn} = \int \mathrm{d}^3 x M_{mno}$. When the action is invariant under conformal transformations the trace of the stress tensor vanishes and the charges D and K_n also become constants of motion. The conformal algebra contains in addition to the Poincaré algebra already given the following commutators

$$[P_l, D] = -\mathrm{i}P_l \qquad [K_l, D] = \mathrm{i}K_l \qquad [M_{lm}, K_n] = \mathrm{i}(\eta_{nl}K_m - \eta_{nm}K_l)$$
$$[P_l, K_m] = 2\mathrm{i}(M_{lm} - \eta_{lm}D) \tag{2.57}$$
$$[M_{lm}, D] = [K_l, K_m] = 0.$$

In fact it is enough to know $[P_n, d]$; the other commutators may then be derived assuming the closure of the algebra, the already known Poincaré algebra and making use of the Jacobi identity.

The superconformal algebra (Wess and Zumino 1974b, Fradkin and Tseytlin 1985) is obtained analogously if we add to the super-Poincaré algebra an additional supersymmetry charge S corresponding to another conserved supercurrent defined from the first moment $s_n = -x^m \gamma_m \hat{J}_n$ of the γ-traceless Majorana supercurrent $\hat{J}_n$ besides the charges D and K_m. The assumption of closure and the graded Jacobi identities require also the axial ($R-$)charge R in the algebra. It cannot be set to zero since $[Q, R] \neq 0$. The algebra with conveniently chosen normalisations reads as follows (apart from (2.57) and (1.75)):

$$[M_{lm}, Q_A] = \mathrm{i}(\gamma_{lm}Q)_A \qquad [M_{lm}, S_A] = \mathrm{i}(\gamma_{lm}S)_A$$
$$[Q, P_l] = [S, K_l] = 0 \qquad [S, P_l] = \gamma_l Q \qquad [Q, K_l] = \gamma_l S$$
$$[Q, D] = \tfrac{1}{2}\mathrm{i}Q \qquad [S, D] = -\tfrac{1}{2}\mathrm{i}S$$
$$[Q, R] = -\mathrm{i}\gamma_5 Q \qquad [S, R] = \mathrm{i}\gamma_5 S \tag{2.58}$$
$$[R, M_{lm}] = [R, P_l] = [R, D] = [R, K_l] = 0$$
$$\{Q_A, \bar{Q}_B\} = -2\gamma^l{}_{AB}P_l \qquad \{S_A, \bar{S}_B\} = -2\gamma^l{}_{AB}K_l$$
$$\{S, \bar{Q}\} = 2\mathrm{i}D + \frac{\mathrm{i}}{2}[\gamma_l, \gamma_m]M^{lm} - 3\mathrm{i}\gamma_5 R.$$

Problems

2.1 Define

$$N_m{}^i{}_j = -\tfrac{1}{2}\bar{\sigma}_m{}^{\dot{\beta}\alpha}[Q_{\alpha j}, \bar{Q}_{\dot{\beta}}{}^i] \qquad N_m = N_m{}^i{}_i$$

and show that

$$[N_l{}^i{}_j, Q_{\beta k}] = 2P^m(\sigma_l\bar{\sigma}_m)_\beta{}^\alpha Q_{\alpha j}\delta^i_k + 4P_l Q_{\beta j}\delta^i_k$$
$$[P\cdot N^i{}_j, Q_{\beta k}] = 2P^2 Q_{\beta j}\delta^i_k$$

and that in the massless case we obtain

$$[N_l{}^i{}_j, Q_{\beta k}] = 4P_l Q_{\beta j}\delta^i_k \qquad P\cdot N^i{}_j = 0.$$

2.2 Show that for the massive case

$$F^i{}_j + \frac{1}{2P^2}P^l(N_l{}^i{}_j - \frac{1}{N}N_l\delta^i_j)$$

commutes with SS generators and constitutes SS invariant generalisation of SU(N) generators while for the massless case the corresponding generalisation is given by

$$W_m{}^i{}_j = P_m F^i{}_j + \tfrac{1}{4}(N_m{}^i{}_j - N^{-1}N_m\delta^i_j)$$

and we may find a matrix representation such that $W_m{}^i{}_j = g^i{}_j P_m$.

2.3 Rewrite the Wess–Zumino Lagrangian in the four-component notation in terms of fields (A, χ_L, F) where χ_L is the left-handed projection of a Majorana spinor field χ.

2.4 Rewrite (problem 2.3) also the supersymmetry and R-currents and the supersymmetry transformations. Derive the SS variation of the supercurrent and show that it leads to the energy–momentum tensor θ_{lm}.

2.5 Show that a conserved symmetric stress tensor, conserved Majorana vector–spinor and conserved ($R-$) axial vector can be constructed for the massless free Wess–Zumino action such that they form a multiplet of currents under SS variations.

2.6 Consider an infinitesimal gauge variation of the Lagrangian $\mathscr{L}[\varphi, \partial\sigma]$ where φ is a field multiplet and $\delta\varphi \approx \varphi'(x) - \varphi(x)$. Show that on the mass-shell $\delta\mathscr{L} \stackrel{\circ}{=} \partial_l\left[\delta\varphi\,\dfrac{\partial\mathscr{L}}{\partial(\partial_l\varphi)}\right]$ and if the action is invariant under the variations with constant parameters ε, viz, $\delta\mathscr{L} = \varepsilon\partial_l\Lambda^l(x)$, the conserved Noether current is given by $\varepsilon j^l_N = \left[\delta\varphi\,\dfrac{\partial\mathscr{L}}{\partial(\partial_l\varphi)} - \varepsilon\Lambda^l\right]$. When the global symmetry is promoted to a local one with space–time dependent parameters show that the variation of the Lagrangian now contains another term proportional to the Noether current, i.e., $\delta\mathscr{L} \overset{\circ}{=} \partial_l(\varepsilon\Lambda^l) + (\partial_l\varepsilon)j_N^l(x) + \ldots$.

2.7 Derive the expressions for the super and axial ($R-$) currents for the Wess–Zumino model.

2.8 The following nonlinear realisation of supersymmetry was given by Volkov and Akulov:

$$\delta\lambda_\alpha = d\xi_\alpha - \frac{\mathrm{i}}{d}(\lambda\sigma^l\bar{\xi} - \xi\sigma^l\bar{\lambda})\partial_l\lambda_\alpha$$

where d is a dimensional constant representing the scale of supersymmetry breaking. Show that this transformation closes into the SS algebra. If ρ is a field with the homogeneous transformation law $\delta\rho = -(\mathrm{i}/d)(\bar{\xi}\gamma^l\lambda)\partial_l\rho$ show that the algebra closes on it as well. Furthermore show that the same holds for the simpler transformation law

$$\delta\lambda_\alpha = -\frac{2\mathrm{i}}{d}(\lambda\sigma^l\bar{\xi})\partial_l\lambda_\alpha + d\xi_\alpha.$$

2.9 Write

$$W_l{}^m = \left[\delta_l{}^m - \frac{\mathrm{i}}{d^2}\bar{\lambda}\gamma^l\partial_l\lambda\right]$$

and show that det **W** transforms as a total divergence under SS transformations and the action for the V–A field may be defined by

$$\mathscr{L}_\lambda = -\frac{d^2}{2}\det\left(\delta_l{}^m - \frac{\mathrm{i}}{d^2}\bar{\lambda}\gamma^l\partial_l\lambda\right).$$

Find the Noether current and verify that the field is massless.

3

SUPERSPACE AND SUPERFIELDS

3.1 Introduction

The formulation of a supersymmetric field theory requires, as seen in Chapter 2, a supermultiplet containing bosonic as well as fermionic fields. The generators of the super-Poincaré group P_l, Q_α, $\bar{Q}_{\dot{\alpha}}$, M_{lm} must be expressed as conserved charges derived from local Noether currents. The algebra of the generators then follows as a direct consequence of the field commutators.

A very useful tool for dealing with supersymmetric field theories is to represent a supermultiplet by a superfield defined on a superspace of co-ordinates spanned by

$$z^M : (x^m, \theta^\alpha, \bar{\theta}_{\dot{\alpha}}) \tag{3.1}$$

where the index M runs over bosonic indices m and fermionic indices α and $\dot{\alpha}$ (Salam and Strathdee 1974a, Ferrara *et al*. 1974, de Witt 1984). The commutation relations among the superspace coordinates are given by†

$$z^M z^N = (-1)^{mn} z^N z^M \tag{3.2}$$

where $m = 0$ or 1 according as M is a bosonic or a fermionic index. When both of the indices M and N are fermionic the corresponding coordinates anticommute. Superfields may be constructed to carry representations of the super-Poincaré group analogous to the case of ordinary fields defined over space–time and carrying a representation of the Poincaré group.

3.2 Coordinate space representation of super-Poincaré generators

We may represent a supersymmetry transformation as a transformation over points in superspace. The well known procedure utilises the fact that a group G may be realised as a transformation group over a coset space G/H where H is a closed subgroup. For example, in the case of Poincaré group $\mathscr{P}$ the coset decomposition with respect to the Lorentz sub-group L defines the coset space $\mathscr{P}/L = \{(a_l, I)L\}$ where $\mathscr{P} = \{(a_l, \Lambda) \mid (a_l', \Lambda')(a_l, \Lambda) =$

† The Grassmann product hence is defined to satisfy $dz^M \wedge dz^N = -(-1)^{mn} dz^N \wedge dz^M$ and $(dz^M) z^N = (-1)^{mn} z^N dz^M$.

$(a'_l + (\Lambda' a)_l, \Lambda'\Lambda);\ \Lambda \in L\}$. Considering the action of an arbitrary element (a_l, Λ) on a fixed element $(x_l, I)L$ of the coset space we obtain

$$(a, \Lambda)(x, I)L = (a + \Lambda x, I)L \equiv (x', I)L. \tag{3.3}$$

In other words we may represent (a, Λ) by the transformation $x'^l = a^l + \Lambda^l{}_m x^m$ over the space–time coordinates which parametrise the coset space. In the case of the super-Poincaré group it is convenient to use an exponential parametrisation of group elements, viz. $\exp\{\mathrm{i}[-a \cdot P + \xi Q + \bar{\xi}\bar{Q} + \frac{1}{2}\lambda^{lm} M_{lm}]\}$ and take the fixed element of the coset space with respect to the Lorentz subgroup to be

$$g_{\mathrm{L}}(x, \theta, \bar{\theta}) = (\exp \mathrm{i}[-x \cdot P + \theta Q + \bar{\theta}\bar{Q}])L. \tag{3.4}$$

On making use of the algebra of $(N = 1)$ supersymmetry generators

$$\begin{aligned} \{Q_\alpha, \bar{Q}_{\dot{\beta}}\} &= 2\sigma^l{}_{\alpha\dot{\beta}} P_l \\ \{Q_\alpha, Q_\beta\} &= \{\bar{Q}_{\dot{\alpha}}, \bar{Q}_{\dot{\beta}}\} = [Q_\alpha, P_l] = [\bar{Q}_{\dot{\alpha}}, P_l] = 0 \end{aligned} \tag{3.5}$$

and the Baker–Hausdorf–Campbell theorems we easily derive

$$\exp \mathrm{i}[-a \cdot P + \xi Q + \bar{\xi}\bar{Q}]\, g_{\mathrm{L}}(x, \theta, \bar{\theta}) = g_{\mathrm{L}}(x', \theta', \bar{\theta}') \tag{3.6}$$

where†

$$\begin{aligned} x'^l &= x^l + \mathrm{i}(\theta\sigma^l\bar{\xi} - \xi\sigma^l\bar{\theta}) + a^l \\ \theta'_\alpha &= \theta_\alpha + \xi_\alpha \\ \bar{\theta}'_{\dot{\alpha}} &= \bar{\theta}_{\dot{\alpha}} + \bar{\xi}_{\dot{\alpha}}. \end{aligned} \tag{3.7}$$

The supersymmetry transformations are thus realised as supertranslations over points in superspace. We remark that x'_l is real if x_l is so. However, $\mathrm{i}(\theta\sigma^l\bar{\xi} - \xi\sigma^l\bar{\theta})$ is a four-vector with nilpotent numbers.

We may as well have done coset decomposition with L on the left and used the right multiplication on the fixed element $L \exp \mathrm{i}(-x \cdot P + \theta Q + \bar{\theta}\bar{Q}) = g_{\mathrm{R}}(x, \theta, \bar{\theta})$:

$$g_{\mathrm{R}}(x, \theta, \bar{\theta}) \exp \mathrm{i}(\xi Q + \bar{\xi}\bar{Q}) = g_{\mathrm{R}}(x'', \theta'', \bar{\theta}''). \tag{3.8}$$

This gives the following supertranslations:

$$\begin{aligned} x''^l &= x^l - \mathrm{i}(\theta\sigma^l\bar{\xi} - \xi\sigma^l\bar{\theta}) \\ \theta''_\alpha &= \theta_\alpha + \xi_\alpha \\ \bar{\theta}''_{\dot{\alpha}} &= \bar{\theta}_{\dot{\alpha}} + \bar{\xi}_{\dot{\alpha}}. \end{aligned} \tag{3.9}$$

In what follows we shall adopt (3.7) as supersymmetry transformations.

† On acting on g_{L} by $\exp(\mathrm{i}/2)\lambda^{lm} M_{lm}$ we simply obtain Lorentz transformations of x^l and θ_α, $\bar{\theta}_{\dot{\alpha}}$. Note that $[\theta] = [\bar{\theta}] = -\frac{1}{2}$.

The differential generators are then defined through

$$\delta z^M = \mathrm{i}(a \cdot P + \xi Q + \bar{\xi}\bar{Q})z^M. \tag{3.10}$$

From (3.7) and $\partial_l x^m = \delta_l{}^m$, $\partial_\alpha \theta^\beta = \delta_\alpha{}^\beta$, $\bar{\partial}_{\dot{\alpha}}\bar{\theta}^{\dot{\beta}} = \delta_{\dot{\alpha}}^{\dot{\beta}}$, $\partial_\alpha = -\varepsilon_{\alpha\beta}\partial^\beta$, $\partial^\alpha = -\varepsilon^{\alpha\beta}\partial_\beta$ etc it follows that

$$\begin{aligned} \mathrm{i}Q_\alpha &= [\partial_\alpha - \mathrm{i}\sigma^l_{\alpha\dot{\beta}}\bar{\theta}^{\dot{\beta}}\partial_l] \\ \mathrm{i}\bar{Q}^{\dot{\alpha}} &= [\bar{\partial}^{\dot{\alpha}} + \mathrm{i}\theta^\alpha\sigma^{l\dot{\alpha}}_\alpha\partial_l] \\ P_l &= -\mathrm{i}\partial_l \end{aligned} \tag{3.11}$$

while $\mathrm{i}Q^\alpha = \mathrm{i}\varepsilon^{\alpha\beta}Q_\beta = [-\partial^\alpha - \mathrm{i}(\sigma^l\bar{\theta})^\alpha\partial_l]$ and $\mathrm{i}\bar{Q}_{\dot{\alpha}} = [-\bar{\partial}_{\dot{\alpha}} + \mathrm{i}(\theta\sigma^l)_{\dot{\alpha}}\partial_l]$. These expressions satisfy (3.5) as expected. The R-symmetry operator is clearly represented by $R = (\theta\partial - \bar{\theta}\bar{\partial})$ and $\delta\theta = \mathrm{i}\alpha R\theta$, $\delta\bar{\theta} = \mathrm{i}\alpha R\bar{\theta}$.

3.3 Superfield, covariant spinorial derivatives

3.3.1 Scalar superfield

A superfield† $F(z) \equiv F(x, \theta, \bar{\theta})$ is a mapping from points in superspace to 'complex' numbers. The Taylor expansion in θ, $\bar{\theta}$ terminates after a finite number of terms and has the general form

$$\begin{aligned} f(z) = f(x) &+ \theta\varphi(x) + \bar{\theta}\bar{\chi}(x) + \theta^2 m(x) + \bar{\theta}^2 n(x) + \theta\sigma^l\bar{\theta}v_l(x) \\ &+ \theta^2\bar{\theta}\bar{\lambda}(x) + \bar{\theta}^2\theta\psi(x) + \theta^2\bar{\theta}^2 d(x). \end{aligned} \tag{3.12}$$

Here $(f, \varphi, \bar{\chi} \ldots d)$ (x) denote component fields of the supermultiplet corresponding to the superfield under discussion. As regards the dimensions of the component fields we note that $[F] = [f(x)]$, $[\varphi(x)] = [f(x)] + \frac{1}{2}$, $\ldots$, $[d(x)] = [f(x)] + 2$. The coordinates θ, $\bar{\theta}$ are book-keeping parameters for the components of a supermultiplet whose infinitesimal variations are defined in terms of that of the superfield by

$$\delta F = \delta f(x) + \theta\, \delta\varphi(x) + \ldots + \theta^2\bar{\theta}^2\, \delta d(x) \tag{3.13}$$

and consequently

$$[\delta_1, \delta_2]F = [\delta_1, \delta_2]f + \theta[\delta_1, \delta_2]\varphi + \ldots. \tag{3.14}$$

The sum and product of two superfields is again a superfield. The form of F at an infinitesimally close point $z + \delta z$ is given by

$$\begin{aligned} F(z + \delta z) &= F(z) + \delta z^M \partial_M F(z) + \ldots \\ &= F(z) + (\delta x^l)\partial_l F + (\delta\theta^\alpha)\partial_\alpha F + (\delta\bar{\theta}_{\dot{\alpha}})\bar{\partial}^{\dot{\alpha}} F + \ldots. \end{aligned} \tag{3.15}$$

† F may in general carry internal or space–time tensorial and spinor indices.

Under a supersymmetry transformation (supertranslation) $z \to z'$, a superfield $F(z)$ is mapped onto $F'(z')$. A complex scalar superfield $S(z)$ may be defined by requiring that it remains invariant under supertranslations

$$S'(z') = S(z). \tag{3.16}$$

For infinitesimal transformations in view of (3.7) and (3.11) we obtain the following SS transformations for $S(z)$:

$$\begin{aligned}\delta S(z) = S'(z) - S(z) &= -(\delta z^M)\partial_M S + \ldots \\ &= -\mathrm{i}(a \cdot P + \xi Q + \bar{\xi}\bar{Q})S + \ldots \\ [\delta_2, \delta_1] S(z) &= 2\mathrm{i}(\xi_2 \sigma^l \bar{\xi}_1 - \xi_1 \sigma^l \bar{\xi}_2) S(z).\end{aligned} \tag{3.17}$$

3.3.2 Covariant spinorial derivatives

From the fact that ∂_l commutes with Q_α, $\bar{Q}_{\dot\alpha}$

$$\delta(\partial_l S) = \partial_l \delta S = -\mathrm{i}(\xi Q + \bar{\xi}\bar{Q})\, \partial_l S \tag{3.18}$$

for an infinitesimal SS transformation it shows that ∂_l is a covariant derivative with respect to these transformations. On the other hand ∂_α, $\bar{\partial}_{\dot\alpha}$ do not define covariant derivatives of S. The covariant spinorial derivatives D_α, $\bar{\mathrm{D}}_{\dot\alpha}$ must satisfy

$$\begin{aligned}\delta \mathrm{D}_\alpha S &= -\mathrm{i}(\xi Q + \bar{\xi}\bar{Q})\mathrm{D}_\alpha S + \ldots \\ &= -\mathrm{i}\mathrm{D}_\alpha(\xi Q + \bar{\xi}\bar{Q})S + \ldots.\end{aligned} \tag{3.19}$$

In view of the explicit expressions for Q_α, $\bar{Q}_{\dot\alpha}$ in (3.11) we may define them to be

$$\begin{aligned}\mathrm{D}_\alpha &= \partial_\alpha + \mathrm{i}(\sigma^l \bar{\theta})_\alpha \partial_l & \bar{\mathrm{D}}_{\dot\alpha} &= -\bar{\partial}_{\dot\alpha} - \mathrm{i}(\theta \sigma^l)_{\dot\alpha} \partial_l \\ \mathrm{D}^\alpha = \varepsilon^{\alpha\beta}\mathrm{D}_\beta &= -\partial^\alpha + \mathrm{i}\sigma^{l\alpha}{}_{\dot\beta}\bar{\theta}^{\dot\beta}\partial_l & \bar{\mathrm{D}}^{\dot\alpha} = \varepsilon^{\dot\alpha\dot\beta}\bar{\mathrm{D}}_{\dot\beta} &= \bar{\partial}^{\dot\alpha} - \mathrm{i}\theta^\alpha \sigma^l{}_\alpha{}^{\dot\alpha}\partial_l.\end{aligned} \tag{3.20}$$

The covariant derivatives are required to impose SS covariant constraints on superfields as well as in the construction of Lagrangians invariant under SS transformations. The following derivative properties may easily be verified

$$\mathrm{D}_\alpha(FG) = (\mathrm{D}_\alpha F)G \pm F(\mathrm{D}_\alpha G) \tag{3.21}$$

where the upper (lower) sign holds according as F is an even (odd) Grassmann function. It follows then

$$\mathrm{D}^2(FG) = \mathrm{D}^\alpha \mathrm{D}_\alpha(FG) = (\mathrm{D}^2 F)G + F(\mathrm{D}^2 G) \pm 2(\mathrm{D}^\alpha F)(\mathrm{D}_\alpha G) \tag{3.22}$$

and analogously

$$\bar{\mathrm{D}}^2(FG) = \bar{\mathrm{D}}_{\dot\alpha}\bar{\mathrm{D}}^{\dot\alpha}(FG) = (\bar{\mathrm{D}}^2 F)G + F(\bar{\mathrm{D}}^2 G) \pm 2(\bar{\mathrm{D}}_{\dot\alpha} F)(\bar{\mathrm{D}}^{\dot\alpha} G). \tag{3.23}$$

3.3.3 *Algebra of* D, $\bar{\mathrm{D}}$ *Operators*

The (anti-) commutation relations

$$\begin{aligned} \{\mathrm{D}_\alpha, \bar{\mathrm{D}}_{\dot\beta}\} &= -2\mathrm{i}\sigma^l{}_{\alpha\dot\beta}\partial_l = 2\sigma^l{}_{\alpha\dot\beta}P_l \\ \{\mathrm{D}_\alpha, \mathrm{D}_\beta\} &= \{\bar{\mathrm{D}}_{\dot\alpha}, \bar{\mathrm{D}}_{\dot\beta}\} = 0 \\ \{\mathrm{D}_\alpha, Q_\beta\} &= \{\bar{\mathrm{D}}_{\dot\alpha}, Q_\beta\} = \{\mathrm{D}_\alpha, \bar{Q}_{\dot\beta}\} = \{\bar{\mathrm{D}}_{\dot\alpha}, \bar{Q}_{\dot\beta}\} = 0 \end{aligned} \tag{3.24}$$

are easily established and we may derive the following frequently used relations

$$\begin{aligned} &\mathrm{D}^\alpha\mathrm{D}^\beta = -\tfrac{1}{2}\varepsilon^{\alpha\beta}\mathrm{D}^2 \qquad \bar{\mathrm{D}}^{\dot\alpha}\bar{\mathrm{D}}^{\dot\beta} = \tfrac{1}{2}\varepsilon^{\dot\alpha\dot\beta}\bar{\mathrm{D}}^2 \\ &\mathrm{D}_\alpha\mathrm{D}_\beta\mathrm{D}_\lambda = \bar{\mathrm{D}}_{\dot\alpha}\bar{\mathrm{D}}_{\dot\beta}\bar{\mathrm{D}}_{\dot\lambda} = 0 \\ &\tfrac{1}{2}[\mathrm{D}_\alpha, \bar{\mathrm{D}}_{\dot\alpha}] = \mathrm{D}_\alpha\bar{\mathrm{D}}_{\dot\alpha} + \mathrm{i}\partial_{\alpha\dot\alpha} = -\bar{\mathrm{D}}_{\dot\alpha}\mathrm{D}_\alpha - \mathrm{i}\partial_{\alpha\dot\alpha} \\ &\tfrac{1}{4}[\mathrm{D}_\alpha, \bar{\mathrm{D}}^2] = -\mathrm{i}\partial_{\alpha\dot\alpha}\bar{\mathrm{D}}^{\dot\alpha} \qquad \tfrac{1}{4}[\bar{\mathrm{D}}_{\dot\alpha}, \mathrm{D}^2] = \mathrm{i}\mathrm{D}^\alpha\partial_{\alpha\dot\alpha} \\ &\tfrac{1}{8}[\mathrm{D}^2, \bar{\mathrm{D}}^2] = -\mathrm{i}\mathrm{D}^\alpha\partial_{\alpha\dot\alpha}\bar{\mathrm{D}}^{\dot\alpha} - 2\Box = \mathrm{i}\bar{\mathrm{D}}^{\dot\alpha}\partial_{\alpha\dot\alpha}\mathrm{D}^\alpha + 2\Box \\ &\mathrm{D}^\alpha\bar{\mathrm{D}}^2\mathrm{D}_\alpha = \bar{\mathrm{D}}_{\dot\alpha}\mathrm{D}^2\bar{\mathrm{D}}^{\dot\alpha} \\ &\mathrm{D}^2\bar{\mathrm{D}}^2\mathrm{D}^2 = 16\Box\mathrm{D}^2 \\ &\bar{\mathrm{D}}^2\mathrm{D}^2\bar{\mathrm{D}}^2 = 16\Box\bar{\mathrm{D}}^2 \end{aligned} \tag{3.25}$$

where

$$\begin{aligned} \partial_{\alpha\dot\alpha} &= \sigma^l{}_{\alpha\dot\alpha}\partial_l \\ \mathrm{D}^2 = \mathrm{D}^\alpha\mathrm{D}_\alpha &= -[\partial^\alpha\partial_\alpha - 2\mathrm{i}\bar\theta^{\dot\beta}\partial_{\alpha\dot\beta}\partial^\alpha + \bar\theta^2\Box] \\ \bar{\mathrm{D}}^2 = \bar{\mathrm{D}}_{\dot\alpha}\bar{\mathrm{D}}^{\dot\alpha} &= -[\bar\partial_{\dot\alpha}\bar\partial^{\dot\alpha} + 2\mathrm{i}\theta^\alpha\partial_{\alpha\dot\beta}\bar\partial^{\dot\beta} + \theta^2\Box] \end{aligned} \tag{3.26}$$

The following projection operators

$$\begin{aligned} P_1 &= \frac{\mathrm{D}^2\bar{\mathrm{D}}^2}{16\Box} \qquad P_2 = \frac{\bar{\mathrm{D}}^2\mathrm{D}^2}{16\Box} \\ -P_\mathrm{T} &= \frac{1}{8\Box}\mathrm{D}^\alpha\bar{\mathrm{D}}^2\mathrm{D}_\alpha = \frac{1}{8\Box}\bar{\mathrm{D}}_{\dot\alpha}\mathrm{D}^2\bar{\mathrm{D}}^{\dot\alpha} \end{aligned} \tag{3.27}$$

satisfying $P_1 + P_2 + P_\mathrm{T} = 1$ together with the operators $P_+ = 1/(4\sqrt{\Box})\mathrm{D}^2$, $P_- = 1/(4\sqrt{\Box})\bar{\mathrm{D}}^2$ form a set of differential operators which is closed under multiplication. It is sometimes convenient to write

$$P_\mathrm{L} = (P_1 + P_2) = \frac{1}{16\Box}(\mathrm{D}^2 + \bar{\mathrm{D}}^2)^2 \tag{3.28}$$

which satisfies $P_\mathrm{L}^2 = P_\mathrm{L}, P_\mathrm{L}P_\mathrm{T} = 0$. The projection operators will be used to define irreducible pieces of a general superfield.

3.3.4 Operator identities

We may greatly simplify the manipulation of the D, $\bar{\mathrm{D}}$ operators by making use of the operator identities derived below (Srivastava 1975):

$$U_b \equiv \mathrm{e}^{\mathrm{i}b(\theta\sigma^l\bar{\theta})\partial_l} = 1 + \mathrm{i}b(\theta\sigma^l\bar{\theta})\partial_l + \frac{b^2}{4}\theta^2\bar{\theta}^2\Box$$

$$\begin{aligned} U_b^{-1}\mathrm{D}_\alpha U_b &= \partial_\alpha + \mathrm{i}(b+1)(\sigma^l\bar{\theta})_\alpha\partial_l \\ U_b^{-1}\bar{\mathrm{D}}_{\dot{\alpha}}U_b &= -\bar{\partial}_{\dot{\alpha}} + \mathrm{i}(b-1)(\theta\sigma^l)_{\dot{\alpha}}\partial_l. \end{aligned} \tag{3.29}$$

Calling $U = \exp[\mathrm{i}(\theta\sigma^l\bar{\theta})\partial_l]$ we find†

$$\begin{aligned} U^{-1}\bar{\mathrm{D}}_{\dot{\alpha}}U &= -\bar{\partial}_{\dot{\alpha}} \\ U^{-1}\mathrm{D}_\alpha U &= \partial_\alpha + 2\mathrm{i}(\sigma^l\bar{\theta})_\alpha\partial_l \end{aligned} \tag{3.30}$$

while

$$\begin{aligned} U\mathrm{D}_\alpha U^{-1} &= \partial_\alpha \\ U\bar{\mathrm{D}}_{\dot{\alpha}}U^{-1} &= -\bar{\partial}_{\dot{\alpha}} - 2\mathrm{i}(\theta\sigma^l)_{\dot{\alpha}}\partial_l. \end{aligned} \tag{3.31}$$

Analogously, we find for the generators

$$\begin{aligned} U\mathrm{i}\bar{Q}_{\dot{\alpha}}U^{-1} &= -\bar{\partial}_{\dot{\alpha}} \\ U\mathrm{i}Q_\alpha U^{-1} &= \partial_\alpha - 2\mathrm{i}(\sigma^l\bar{\theta})_\alpha\partial_l \end{aligned} \tag{3.32}$$

and

$$\begin{aligned} U^{-1}\mathrm{i}Q_\alpha U &= \partial_\alpha \\ U^{-1}\mathrm{i}\bar{Q}_{\dot{\alpha}}U &= -\bar{\partial}_{\dot{\alpha}} + 2\mathrm{i}(\theta\sigma^l)_{\dot{\alpha}}\partial_l. \end{aligned} \tag{3.33}$$

It is easy to verify that

$$U_b^{-1}(x^l, \theta_\alpha, \bar{\theta}_{\dot{\alpha}})U_b = (x^l - \mathrm{i}b\theta\sigma^l\bar{\theta}, \theta_\alpha, \bar{\theta}_{\dot{\alpha}})$$

$$U_b^{-1}\partial_l U_b = \partial_l \qquad U^{-1}(\theta\partial - \bar{\theta}\bar{\partial})U = (\theta\partial - \bar{\theta}\bar{\partial})$$

and consequently

$$\bar{\mathrm{D}}_{\dot{\alpha}}y^l = \mathrm{D}_\alpha y^{+l} = 0 \tag{3.34}$$

where $y^l = x^l + \mathrm{i}\theta\sigma^l\bar{\theta} = Ux^l$ and $y^{+l} = x^l - \mathrm{i}\theta\sigma^l\bar{\theta} = U^{-1}x^l$. The action of U_b on a superfield $F(z) = F(x, \theta, \bar{\theta})$ is a translation in x^l

$$\begin{aligned} U_bF(x,\theta,\bar{\theta}) &= F(x^l + \mathrm{i}b\theta\sigma^l\bar{\theta}, \theta, \bar{\theta}) \\ U_b[F(z)G(z)] &= [U_bF(z)][U_bG(z)]. \end{aligned} \tag{3.35}$$

† $U^{-1}\bar{\mathrm{D}}^{\dot{\alpha}}U = \bar{\partial}^{\dot{\alpha}}$, $\qquad U\mathrm{D}^\alpha U^{-1} = -\partial^\alpha$ etc.

3.4 Irreducible scalar superfield

Under an SS transformation the component fields of a scalar supermultiplet (or superfield) transform, according to (3.17), among themselves linearly. An irreducible supermultiplet must contain at least one fermion and one boson field. We may use the covariant projection operators P_1, P_2, P_T discussed in the previous section to construct the following irreducible components of a general scalar superfield†

$\Phi(x,\theta,\bar{\theta}) = P_2 S(x,\theta,\bar{\theta})$ chiral scalar superfield

$\bar{\Phi}(x,\theta,\bar{\theta}) = P_1 S(x,\theta,\bar{\theta})$ anti-chiral scalar

$G(x,\theta,\bar{\theta}) = P_T S(x,\theta,\bar{\theta})$ linear scalar superfield or non-chiral transverse vector multiplet. (3.36)

They may equivalently be defined by imposing the following SS covariant differential constraints

$$\bar{\mathrm{D}}_{\dot{\alpha}}\Phi = 0 \qquad \mathrm{D}_{\alpha}\bar{\Phi} = 0.$$
$$\mathrm{D}^2 G = \bar{\mathrm{D}}^2 G = 0. \tag{3.37}$$

We note an inherent gauge invariance, say, associated with the chiral superfield. Both Φ and $\Phi + \bar{\mathrm{D}}^2\Omega$, where Ω is an arbitrary scalar superfield, satisfy the chiral condition. Φ and $\bar{\Phi}$ carry matter fields and will be discussed in detail in the next chapter. Towards an illustration we derive the component fields contained in $G(x,\theta,\bar{\theta})$. From the operator identities in (3.31) and (3.37) it follows that $\bar{\partial}^2[U^{-1}G] = \partial^2[UG] = 0$. Collecting then the θ^2 component in UG and $\bar{\theta}^2$ component in $U^{-1}G$ and setting them to zero we obtain

$$m + \bar{\theta}\bar{\lambda} + \bar{\theta}^2 d - \frac{\mathrm{i}}{2}(\partial_l \varphi)\sigma^l\bar{\theta} - \frac{\mathrm{i}}{2}\bar{\theta}^2\partial_l v^l + \frac{1}{4}\bar{\theta}^2\Box f = 0$$
$$n + \theta\psi + \theta^2 d + \frac{\mathrm{i}}{2}\theta\sigma^l\partial_l\bar{\chi} + \frac{\mathrm{i}}{2}\theta^2\partial_l v^l + \frac{1}{4}\theta^2\Box f = 0 \tag{3.38}$$

and consequently

$$m = n = 0 \qquad d = -\frac{1}{4}\Box f \qquad \partial_l v^l = 0 \text{ (transverse)}$$
$$\bar{\lambda}_{\dot{\alpha}} = +\frac{\mathrm{i}}{2}\partial_{\alpha\dot{\alpha}}\varphi^{\alpha} \qquad \psi_{\alpha} = -\frac{\mathrm{i}}{2}\partial_{\alpha\dot{\alpha}}\bar{\chi}^{\dot{\alpha}}. \tag{3.39}$$

An irreducible supermultiplet may also be obtained as a submultiplet by

† The component fields of S are $f, \varphi, \ldots, d$ as given in (3.12).

imposing the reality constraint

$$V(x,\theta,\bar\theta)=\bar V(x,\theta,\bar\theta) \tag{3.40}$$

on the scalar superfield. This is the gauge or vector supermultiplet since it carries among its components a vector gauge field. The chiral supermultiplet contains only spin-0 and spin-$\frac{1}{2}$ matter fields.

3.5 Differential forms in flat superspace

From the discussions in §§ 1.2 and 3.1 it is clear that we may introduce a 'flat superspace' metric η_{MN} to lower the index M of the superspace coordinate z^M: $(x^m, \theta^\mu, \bar\theta_{\dot\mu})$. We have

$$\eta_{MN}=\begin{pmatrix}\eta_{mn} & \mathbf{0} & \\ \mathbf{0} & \varepsilon_{\mu\nu} & 0 \\ & 0 & \varepsilon^{\dot\mu\dot\nu}\end{pmatrix}=\begin{pmatrix}\eta_{mn} & \mathbf{0}\\ \mathbf{0} & \mathrm{i}\mathbf{C}\end{pmatrix} \tag{3.41}$$

where $\mathbf{C}$ is given in (1.21). The inverse metric is defined by $\eta_{ML}\eta^{LN}=\delta_M^N$ and we find

$$\eta^{MN}=\begin{pmatrix}\eta^{mn} & \mathbf{0} & \\ \mathbf{0} & \varepsilon^{\mu\nu} & 0 \\ & 0 & \varepsilon_{\dot\mu\dot\nu}\end{pmatrix}=\begin{pmatrix}\eta^{mn} & \mathbf{0}\\ \mathbf{0} & -\mathrm{i}\mathbf{C}^{-1}\end{pmatrix} \tag{3.42}$$

where we define

$$\delta_N^M=\frac{\partial}{\partial z^N}z^M=\begin{pmatrix}\delta_n^m & \mathbf{0} & \\ \mathbf{0} & \delta_\nu^\mu & \\ & & \delta_{\dot\mu}^{\dot\nu}\end{pmatrix} \tag{3.43}$$

and it follows that

$$\frac{\partial}{\partial z^N}z_M=\frac{\partial}{\partial z^N}(\eta_{ML}z^L)=(-1)^{n(m+l)}\eta_{ML}\frac{\partial z^L}{\partial z^N}=(-1)^{n(m+n)}\eta_{MN}=\eta_{MN}$$

and we find similarly

$$\partial^N z^M=\eta^{NM}\qquad \partial^N z_M=\delta_M^N.$$

Under the super-Lorentz transformation $z_M \rightarrow z'_M = \Lambda_M{}^L z_L$ where

$$(\Lambda_M{}^L) = \begin{pmatrix} \Lambda_m{}^l & & \mathbf{0} \\ \mathbf{0} & S_\mu{}^\nu & 0 \\ & 0 & S^{-1\dagger\dot{\mu}}{}_{\dot{\nu}} \end{pmatrix} \tag{3.44}$$

$x^M y_M = (x^m y_m + \theta\eta + \bar{\theta}\bar{\eta})$ is left invariant. From $z^M \eta_{MN} z^N = (-1)^{m(m+2n)}(-1)^{n(m+n)} \; z^N \eta_{MN} z^M = z^N \eta_{NM} z^M$ we derive the graded symmetry property $\eta_{MN} = (-1)^{m+n+mn} \eta_{NM}$ of the metric tensor which is also true for the case of curved superspace.

Under the general supercoordinate transformations $z^{M'} = z^{M'}(z)$ we obtain

$$\mathrm{d}z^{M'} = \mathrm{d}z^N \frac{\partial z^{M'}}{\partial z^N} \tag{3.45}$$

where we keep the differentials $\mathrm{d}z^M$ on the left of their coefficients. For example,

$$\mathrm{d}\bar{\theta}'_{\dot{\mu}} = \mathrm{d}x^m \frac{\partial \bar{\theta}'_{\dot{\mu}}}{\partial x^m} + \mathrm{d}\theta^\mu \frac{\partial \bar{\theta}'_{\dot{\mu}}}{\partial \theta^\mu} + \mathrm{d}\bar{\theta}_{\dot{\nu}} \frac{\partial \bar{\theta}_{\dot{\mu}}}{\partial \bar{\theta}_{\dot{\nu}}}. \tag{3.46}$$

We also have

$$\begin{aligned} &\frac{\partial}{\partial z^{M'}} = \frac{\partial z^N}{\partial z^{M'}} \frac{\partial}{\partial z^N} & &\frac{\partial}{\partial z^M} = \frac{\partial z'^N}{\partial z^M} \frac{\partial}{\partial z'^N} \\ &\delta^L_M = \frac{\partial z^L}{\partial z^M} = \frac{\partial z'^N}{\partial z^M} \frac{\partial z^L}{\partial z'^N} & &\delta^L_M = \frac{\partial z^N}{\partial z^{M'}} \frac{\partial z^{L'}}{\partial z^N} \end{aligned} \tag{3.47}$$

and we derive $\mathrm{d}' = \mathrm{d}$ where $\mathrm{d} = \mathrm{d}z^M \partial/\partial z^M$ is the exterior derivation operator which is useful in discussing differential forms on the superspace.

Under the rigid SS transformations the spinorial derivatives ∂_μ, $\bar{\partial}_{\dot{\mu}}$ do not map superfields into superfields. They do not (anti-) commute with SS generators. It is therefore convenient not to use the natural basis $(\mathrm{d}z^M, \partial/\partial z^M)$ and to introduce the SS covariant basis (e^A, D_A) defined below. The advantage of the description of flat superspace theories in terms of differential forms is that it generalises easily to curved superspace where supergravity makes its appearance.

We note that under the SS transformations $\delta x^m = \mathrm{i}(\theta\sigma^m\bar{\xi} - \xi\sigma^m\bar{\theta})$, $\delta\theta_\mu = \xi_\mu$, $\delta\bar{\theta}_{\dot{\mu}} = \bar{\xi}_{\dot{\mu}}$ the differentials $\mathrm{d}z^M$ transform as

$$\mathrm{d}z^{M'} = (\mathrm{d}x^m + \mathrm{i}\,[\mathrm{d}\theta\sigma^m\bar{\xi} - \xi\sigma^m\,\mathrm{d}\bar{\theta}],\, \mathrm{d}\theta^\mu,\, \mathrm{d}\bar{\theta}_{\dot{\mu}}). \tag{3.48}$$

We may thus construct the following SS invariant linear combinations

$$\begin{aligned} e^a &= (\mathrm{d}x^m)\,\delta^a_m - \mathrm{i}(\mathrm{d}\theta^\mu \sigma^a_{\mu\dot{\mu}} \bar{\theta}^{\dot{\mu}} - \theta^\mu \sigma^a_{\mu\dot{\mu}}\,\mathrm{d}\bar{\theta}^{\dot{\mu}}) \\ e^\alpha &= \mathrm{d}\theta^\alpha = (\mathrm{d}\theta^\mu)\,\delta^\alpha_\mu \\ e_{\dot{\alpha}} &= \mathrm{d}\bar{\theta}_{\dot{\alpha}} = (\mathrm{d}\bar{\theta}_{\dot{\mu}})\,\delta^{\dot{\mu}}_{\dot{\alpha}} \end{aligned} \tag{3.49}$$

or

$$e^A(z) = \mathrm{d}z^M e_M{}^A(z) = \mathrm{d}x^m e_m{}^A + \mathrm{d}\theta^\mu e_\mu{}^A + \mathrm{d}\bar{\theta}_{\dot{\mu}} e^{\dot{\mu}A} \tag{3.50}$$

where $e^A = (e^a, e^\alpha, e_{\dot{\alpha}})$ and

$$e_M{}^A = \begin{pmatrix} e_m{}^a & e_m{}^\alpha & e_{m\dot{\alpha}} \\ e_\mu{}^a & e_\mu{}^\alpha & e_{\mu\dot{\alpha}} \\ e^{\dot{\mu}a} & e^{\dot{\mu}\alpha} & e^{\dot{\mu}}{}_{\dot{\alpha}} \end{pmatrix} = \begin{pmatrix} \delta^a_m & 0 & 0 \\ -\mathrm{i}(\sigma^a\bar{\theta})_\mu & \delta^\alpha_\mu & 0 \\ -\mathrm{i}(\bar{\sigma}^a\theta)^{\dot{\mu}} & 0 & \delta^{\dot{\mu}}_{\dot{\alpha}} \end{pmatrix} = \begin{pmatrix} \delta^a{}_m & & \mathbf{0} \\ -\gamma^a\theta & \delta^\alpha_\mu & 0 \\ & 0 & \delta^{\dot{\mu}}_{\dot{\alpha}} \end{pmatrix}. \tag{3.51}$$

We may also define $e_A = \eta_{AB}e^B$ where the metric η_{AB} has the same elements as in η_{MN} etc. For $\partial/\partial z^M$ we find

$$\frac{\partial}{\partial x^m} = \frac{\partial z'^N}{\partial x^m}\frac{\partial}{\partial z'^N} = \frac{\partial}{\partial x^{m\prime}}$$

$$\frac{\partial}{\partial\theta^\mu} = \frac{\partial x'^n}{\partial\theta^\mu}\frac{\partial}{\partial x'^n} + \frac{\partial\theta'^\nu}{\partial\theta^\mu}\frac{\partial}{\partial\theta'^\nu} + \frac{\partial\bar{\theta}'_{\dot{\nu}}}{\partial\theta^\mu}\frac{\partial}{\partial\bar{\theta}'_{\dot{\nu}}}$$

$$= \mathrm{i}(\sigma^n\bar{\xi})\frac{\partial}{\partial x'^n} + \frac{\partial}{\partial\theta^{\mu\prime}} = \mathrm{i}\sigma^n(\bar{\theta}' - \bar{\theta})\frac{\partial}{\partial x'^n} + \frac{\partial}{\partial\theta^{\mu\prime}}$$

or

$$\mathrm{D}_\mu = \frac{\partial}{\partial\theta^\mu} + \mathrm{i}(\sigma^n\bar{\theta})_\mu\frac{\partial}{\partial x^n} = \frac{\partial}{\partial\theta^{\mu\prime}} + \mathrm{i}(\sigma^n\bar{\theta}')_\mu\frac{\partial}{\partial x'^n} = \mathrm{D}'_\mu \tag{3.52}$$

and similarly

$$\bar{\mathrm{D}}_{\dot{\mu}} = \bar{\mathrm{D}}'_{\dot{\mu}}. \tag{3.53}$$

The covariant derivatives are invariant derivatives as well. We may thus write the SS invariant derivatives $\mathrm{D}_A = (\partial_a, \mathrm{D}_\alpha, \bar{\mathrm{D}}^{\dot{\alpha}})$ as follows

$$\begin{aligned} \mathrm{D}_\alpha &= \delta^\mu_\alpha\frac{\partial}{\partial\theta^\mu} + \mathrm{i}(\sigma^m\bar{\theta})_\mu\,\delta^\mu_\alpha\frac{\partial}{\partial x^m} \\ \bar{\mathrm{D}}^{\dot{\alpha}} &= \delta^{\dot{\alpha}}_{\dot{\mu}}\frac{\partial}{\partial\bar{\theta}_{\dot{\mu}}} + \mathrm{i}(\bar{\sigma}^m\theta)^{\dot{\mu}}\,\delta^{\dot{\alpha}}_{\dot{\mu}}\frac{\partial}{\partial x^m} \end{aligned} \tag{3.54}$$

and define $(\mathrm{D}_l = \partial_l)$

$$\mathrm{D}_A = e_A(z)^M\frac{\partial}{\partial z^M}$$

where

$$e_A{}^M = \begin{pmatrix} e_a{}^m & e_a{}^\mu & e_{a\dot{\mu}} \\ e_\alpha{}^m & e_\alpha{}^\mu & e_{\alpha\dot{\mu}} \\ e^{\dot{\alpha}m} & e^{\dot{\alpha}\mu} & e^{\dot{\alpha}}{}_{\dot{\mu}} \end{pmatrix} = \begin{pmatrix} \delta_a^m & 0 & 0 \\ \mathrm{i}(\sigma^m\bar{\theta})_\mu\,\delta_\alpha^\mu & \delta_\alpha^\mu & 0 \\ \mathrm{i}(\bar{\sigma}^m\theta)^{\dot{\mu}}\,\delta_{\dot{\mu}}^{\dot{\alpha}} & 0 & \delta_{\dot{\mu}}^{\dot{\alpha}} \end{pmatrix} = \begin{pmatrix} \delta_a^m & & 0 \\ \gamma^m\theta & \delta_\alpha^\mu & 0 \\ & 0 & \delta^{\dot{\alpha}}{}_{\dot{\mu}} \end{pmatrix} \tag{3.55}$$

The derivatives D_A map superfields to another superfield and we note that $\mathrm{D}_A z^M = e_A{}^M$. We have chosen the normalisation such that

$$e_M{}^A e_A{}^N = \delta_M{}^N \tag{3.56}$$

and it follows that

$$e_A{}^N e_N{}^B = \delta_A{}^B$$
$$\mathrm{d} = e^A \mathrm{D}_A = \mathrm{d}z^M \frac{\partial}{\partial z^M}. \tag{3.57}$$

Exterior products in superspace are defined by

$$\mathrm{d}z^M \wedge \mathrm{d}z^N = -(-1)^{mn}\,\mathrm{d}z^N \wedge \mathrm{d}z^M$$
$$(\mathrm{d}z^M)z^N = (-1)^{mn} z^N\,\mathrm{d}z^M \tag{3.58}$$

and a differential p-form by

$$\Omega_p = \mathrm{d}z^{M_1} \wedge \ldots \wedge \mathrm{d}z^{M_p} W_{M_p \ldots M_1}(z). \tag{3.59}$$

Functions of z are zero-forms while, for example, e^A is a 1-form. There is no value of p above which all forms vanish unlike in the ordinary case. The familiar rules of multiplication of forms are reproduced:

$$(c_1\Lambda_1 + c_2\Lambda_2) \wedge \Omega = c_1\Lambda_1 \wedge \Omega + c_2\Lambda_2 \wedge \Omega$$
$$\Lambda_p \wedge \Omega_q = (-1)^{pq}\,\Omega_q \wedge \Lambda_p \tag{3.60}$$
$$(\Lambda_1 \wedge \Omega) \wedge \Sigma = \Lambda_1 \wedge (\Omega \wedge \Sigma).$$

The exterior derivative d in (3.57) takes a p-form to a $(p+1)$-form:

$$\mathrm{d}\Omega_p = \mathrm{d}z^{M_1} \wedge \ldots \wedge \mathrm{d}z^{M_p} \wedge \mathrm{d}z^N \frac{\partial}{\partial z^N} W_{M_p \ldots M_1} \tag{3.61}$$

and satisfies in general

$$\mathrm{d}(\Omega + \Sigma) = \mathrm{d}\Omega + \mathrm{d}\Sigma$$
$$\mathrm{d}(\Omega \wedge \Sigma_q) = \Omega \wedge \mathrm{d}\Sigma_q + (-1)^q\,\mathrm{d}\Omega \wedge \Sigma_q \tag{3.62}$$
$$\mathrm{dd} = 0 \qquad \text{(super-Poincaré lemma).}$$

We note, however, that in the flat superspace basis $\mathrm{d}e^A \neq 0$. In fact,

$$\begin{aligned} \mathrm{d}e^A &= \mathrm{d}z^M \wedge \mathrm{d}z^N \frac{\partial}{\partial z^N} e_M{}^A \\ &= (-2\mathrm{i}e^\alpha \wedge \sigma^a_{\alpha\dot\alpha} e^{\dot\alpha}, 0, 0). \end{aligned} \tag{3.63}$$

We will not pursue the subject further and refer the reader to the reference cited (Wess and Bagger 1983).

Problems

3.1 Show that $[\theta \cdot \partial, \theta_\alpha] = \theta_\alpha$, $[\theta\partial, \partial_\alpha] = -\partial_\alpha$, $[\bar\theta\bar\partial, \bar\theta_{\dot\alpha}] = \bar\theta_{\dot\alpha}$, $[\bar\theta\bar\partial, \bar\partial_{\dot\alpha}] = -\bar\partial_{\dot\alpha}$. It follows from (1.79) and (3.11) that the R generator may be represented on the superspace by $R = (\theta\partial - \bar\theta\bar\partial)$. Verify also $URU^{-1} = R$.

3.2 Prove the properties of the D and $\bar{\mathrm{D}}$ operators in (3.25) and (3.26).

3.3 Use the operator identities in (3.30) and (3.31) in conjunction with the property (Chapter 8), $(\theta - \theta')^2 = \delta^2(\theta - \theta')$, $(\bar\theta - \bar\theta')^2 = \delta^2(\bar\theta - \bar\theta')$ to derive the following

$$\mathrm{D}^2(\theta - \theta')^2 = -4\mathrm{e}^{-\mathrm{i}(\theta - \theta')\sigma \cdot \partial\bar\theta}$$

$$\bar{\mathrm{D}}^2(\bar\theta - \bar\theta')^2 = -4\mathrm{e}^{\mathrm{i}\theta\sigma \cdot \partial(\bar\theta - \bar\theta')}$$

$$\mathrm{D}^2\bar{\mathrm{D}}^2(\theta - \theta')^2(\bar\theta - \bar\theta')^2 = 16\mathrm{e}^{-\mathrm{i}[\theta\sigma^l\bar\theta + \theta'\sigma^l\bar\theta' - 2\theta'\sigma^l\bar\theta]\partial_l}$$

$$\bar{\mathrm{D}}^2\mathrm{D}^2(\theta - \theta')^2(\bar\theta - \bar\theta')^2 = 16\mathrm{e}^{\mathrm{i}[\theta\sigma^l\bar\theta + \theta'\sigma^l\bar\theta' + 2\theta\sigma^l\bar\theta']\partial_l}$$

$$\mathrm{D}^2\mathrm{e}^{\mathrm{i}\theta\sigma^l\bar\theta\partial_l} = -4\bar\theta^2\Box$$

$$\bar{\mathrm{D}}^2\mathrm{e}^{-\mathrm{i}\theta\sigma^l\bar\theta\partial_l} = -4\theta^2\Box$$

where $\sigma \cdot \partial = \sigma^l\partial_l$ and both sides act only on an x-dependent function. Hints:

$$\mathrm{D}^2(\theta - \theta')^2 = -U^{-1}\partial^2\mathrm{e}^{\mathrm{i}\theta\sigma \cdot \partial\bar\theta}(\theta - \theta')^2 = -U^{-1}\mathrm{e}^{\mathrm{i}\theta'\sigma \cdot \partial\bar\theta}\partial^2(\theta - \theta')^2$$

$$\mathrm{D}^2\bar{\mathrm{D}}^2(\theta - \theta')^2(\bar\theta - \bar\theta')^2 = -4\mathrm{e}^{\mathrm{i}\theta'\sigma \cdot \partial(\bar\theta - \bar\theta')}\mathrm{D}^2(\theta - \theta')^2$$

$$\mathrm{D}^2\mathrm{e}^{\mathrm{i}\theta\sigma \cdot \partial\bar\theta} = -U^{-1}\partial^2\mathrm{e}^{2\mathrm{i}\theta\sigma \cdot \partial\bar\theta} = -4U^{-1}\bar\theta^2\Box$$

$$\mathrm{D}^2 = -U^{-1}\partial^2 U = -\{\partial^2 - \mathrm{i}[\theta\sigma \cdot \partial\bar\theta, \partial^2] - \tfrac{1}{2}[\theta\sigma \cdot \partial\bar\theta, [\theta\sigma \cdot \partial\bar\theta, \partial^2]]\}.$$

3.4 Show that $(\mathrm{i}Q_\alpha)^* = \mathrm{i}\bar Q_{\dot\alpha}$, $(\mathrm{i}\bar Q^{\dot\alpha})^* = \mathrm{i}Q^\alpha$, $(\mathrm{D}_\alpha)^* = \bar{\mathrm{D}}_{\dot\alpha}$, $(\bar{\mathrm{D}}^{\dot\alpha})^* = \mathrm{D}^\alpha$. Hints: $(\partial_\alpha)^* = -\bar\partial_{\dot\alpha}$, $[(\sigma_l\bar\theta)_\alpha]^* = (\theta\sigma_l)_{\dot\alpha}$ etc.

3.5 Construct the infinitesimal line element invariant under the (flat-) superspace translations.

3.6 Find the SS transformations of y_l and $y_l^\dagger$. Construct the SS and R generators in terms of the chiral superspace variables (y_l, θ).

3.7 In four-component notation, $\psi^{\mathrm{T}} = (\psi_A)^{\mathrm{T}} \equiv (\psi_1, \psi_2, \psi_3, \psi_4) = (\chi_\alpha, \bar{\chi}_c^{\dot{\alpha}})$, show that the SS generator and covariant derivative takė the form

$$Q = -\mathrm{i}\mathbf{C}\frac{\partial}{\partial\theta} - (\gamma^l\theta)\frac{\partial}{\partial x^l} \qquad \bar{Q} = -Q^{\mathrm{T}}\mathbf{C}^{-1}$$

$$\mathrm{D} = -\mathrm{i}\mathbf{C}\frac{\partial}{\partial\theta} + (\gamma^l\theta)\frac{\partial}{\partial x^l} \qquad \bar{\mathrm{D}} = -\mathrm{D}^{\mathrm{T}}\mathbf{C}^{-1}$$

where $\mathrm{D}^{\mathrm{T}} = (\mathrm{D}_A) = (\mathrm{D}_\alpha, \bar{\mathrm{D}}^{\dot{\alpha}})$. Verify that Q satisfies the SS algebra of problem 1.5, and find the algebra of D, $\bar{\mathrm{D}}$ and the identities analogous to (3.25) and (3.29) using the Majorana representation for the gamma matrices.

3.8 Find the supersymmetry generators in two and three space–time dimensions.

3.9 Derive the SS transformation laws for the component fields of the linear scalar supermultiplet G defined in (3.37). Verify its irreducibility and the closure of SS algebra on its component fields.

3.10 Verify that the set P_1, P_2, P_{T}, P_+, P_- is closed under multiplication.

4

CHIRAL SUPERFIELD

4.1 Chiral and anti-chiral scalar superfields

4.1.1 Chiral supermultiplet

The scalar superfields Φ and $\bar{\Phi}$ contain (quark) matter fields with spin 0 and 1/2. The differential constraints in (3.37) may be easily solved. For a chiral superfield, for example, on making use of the operator identities (3.30)

$$U^{-1}\bar{\mathrm{D}}_{\dot\alpha}UU^{-1}\Phi = -\,\bar{\partial}_{\dot\alpha}(U^{-1}\Phi) = 0 \tag{4.1}$$

we obtain the following general form for Φ ($U = \exp[\mathrm{i}(\theta\sigma^l\bar{\theta}\partial_l)]$):

$$\Phi(x,\theta,\bar{\theta}) = U\Psi(x,\theta) = \Psi(y,\theta) \tag{4.2}$$

where

$$\begin{aligned}\Psi(x,\theta) &= A(x) + \sqrt{2}\,\theta\psi(x) + \theta^2 F(x)\\ y^l &= x^l + \mathrm{i}\theta\sigma^l\bar{\theta}.\end{aligned} \tag{4.3}$$

The chiral supermultiplet thus contains a left-handed Majorana spinor $\psi(x)$, a complex scalar $A(x)$ and an auxiliary complex scalar F whose dimension is one unit higher than that of the (physical) field $A(x)$. We note that $A(y)$, $\psi(y)$, $F(y)$ are themselves superfields. The sum and product of two chiral superfields is again a chiral superfield. We record for future use

$$\begin{aligned}&\Phi_i{:}(A_i(x),\,\psi_i(x),\,F_i(x))\\ &\Phi_i\Phi_j{:}(A_iA_j,\,\psi_iA_j + A_i\psi_j,\,A_iF_j + A_jF_i - \psi_i\psi_j)\\ &\Phi_i\Phi_j\Phi_k{:}(A_iA_jA_k,\,\psi_iA_jA_k + \psi_jA_kA_i + \psi_kA_iA_j,\,F_iA_jA_k +\\ &\qquad F_jA_kA_i + F_kA_iA_j - \psi_i\psi_jA_k - \psi_j\psi_kA_i - \psi_k\psi_iA_j).\end{aligned} \tag{4.4}$$

It is very convenient to express the component fields of a superfield as different covariant derivatives of the superfield evaluated at $\theta = \bar{\theta} = 0$. For the chiral superfield

$$\begin{aligned}A(x) &= \Phi(x,\theta,\bar{\theta})|\\ \sqrt{2}\,\psi_\alpha(x) &= \mathrm{D}_\alpha\Phi(x,\theta,\bar{\theta})|\\ F(x) &= -\tfrac{1}{4}\,\mathrm{D}^2\Phi(x,\theta,\bar{\theta})|\end{aligned} \tag{4.5}$$

where '|' indicates that the quantity is evaluated at $\theta = \bar{\theta} = 0$. The derivation of (4.5) is straightforward if we use (3.30). In fact,

$$\begin{aligned}
\mathrm{D}_\alpha\Phi &= UU^{-1}\mathrm{D}_\alpha U\Psi(x,\theta) = U[\partial_\alpha + 2\mathrm{i}(\sigma^l\bar{\theta})_\alpha\partial_l]\Psi(x,\theta)\\
&= U[\sqrt{2}\,\psi_\alpha(x) + \theta\text{-dependent terms}]\\
\mathrm{D}^2\Phi &= UU^{-1}\mathrm{D}^\alpha UU^{-1}\mathrm{D}_\alpha U\Psi(x,\theta) = U[-\partial^\alpha\partial_\alpha + \theta\text{-dependent terms}]\Psi\\
&= U[-4F(x) + \theta\text{-dependent terms}] \qquad (4.6)
\end{aligned}$$

and (4.5) follows.

The chiral supermultiplet (A, ψ, F) is an irreducible representation of SS algebra since only one spinor field is present. The transformation properties of the component fields under an SS transformation are easily derived from (3.17) and (4.5):

$$\begin{aligned}
\delta A(x) &= \delta\Phi\,| = -\mathrm{i}(\xi Q + \bar{\xi}\bar{Q})\Phi| = -(\xi\mathrm{D} + \bar{\xi}\bar{\mathrm{D}})\Phi\,|\\
&= -\xi\mathrm{D}\Phi\,| = -\sqrt{2}\xi\psi(x)\\
\sqrt{2}\,\delta\psi_\alpha(x) &= \delta\mathrm{D}_\alpha\Phi\,| = -\mathrm{i}(\xi Q + \bar{\xi}\bar{Q})\mathrm{D}_\alpha\Phi\,| = -(\xi\mathrm{D} + \bar{\xi}\bar{\mathrm{D}})\mathrm{D}_\alpha\Phi|\\
&= -\xi^\beta\mathrm{D}_\beta\mathrm{D}_\alpha\Phi\,| + \bar{\xi}^{\dot{\beta}}\{\bar{\mathrm{D}}_{\dot{\beta}}, \mathrm{D}_\alpha\}\Phi\,|\\
&= +\tfrac{1}{2}\,\xi_\alpha\mathrm{D}^2\Phi\,| - 2\mathrm{i}\bar{\xi}^{\dot{\beta}}\sigma^l_{\alpha\dot{\beta}}\partial_l\Phi\,| = -2\xi_\alpha F(x) - 2\mathrm{i}(\sigma^l\bar{\xi})_\alpha\partial_l A(x)\\
-4\delta F(x) &= \delta\mathrm{D}^2\Phi\,| = -(\xi\mathrm{D} + \bar{\xi}\bar{\mathrm{D}})\mathrm{D}^2\Phi\,| = -\bar{\xi}_{\dot{\alpha}}[\bar{\mathrm{D}}^{\dot{\alpha}}, \mathrm{D}^2]\Phi\,|\\
&= 4\mathrm{i}(\sigma^l\bar{\xi})_{\alpha\dot{\alpha}}\mathrm{D}^\alpha\partial_l\Phi\,| = -4\sqrt{2}\,\mathrm{i}(\partial_l\psi)\sigma^l\bar{\xi} \qquad (4.7)
\end{aligned}$$

where we use (3.25) and $\bar{\mathrm{D}}_{\dot{\alpha}}\Phi = 0$. The SS tranformations of component field operators of the chiral supermultiplet are thus †

$$\begin{aligned}
\delta A(x) &= \mathrm{i}[\xi Q, A(x)] = -\sqrt{2}\,\xi\psi(x)\\
\delta\psi_\alpha(x) &= \mathrm{i}[\xi Q + \bar{\xi}\bar{Q}, \psi_\alpha(x)] = -\sqrt{2}\,[\xi_\alpha F(x) + \mathrm{i}(\sigma^l\bar{\xi})_\alpha\partial_l A(x)] \qquad (4.8)\\
\delta F(x) &= \mathrm{i}[\bar{\xi}\bar{Q}, F(x)] = \mathrm{i}\sqrt{2}\,\partial_l\psi(x)\sigma^l\bar{\xi}.
\end{aligned}$$

Clearly

$$[\bar{Q}_{\dot{\alpha}}, A(x)] = [Q_\alpha, F(x)] = 0 \qquad \sqrt{2}\,\partial_{\alpha\dot{\alpha}}A = \{\bar{Q}_{\dot{\alpha}}, \psi_\alpha\} \text{ etc} \qquad (4.9)$$

and the coefficient of θ^2—the component $F(x)$—with highest dimension transforms into a total divergence. This observation will be useful for constructing SS invariants.

We note for future applications (use (3.33))

$$\begin{aligned}
\delta\Phi(x,\theta,\bar{\theta}) &= U\delta\Psi(x,\theta)\\
\delta\Psi(x,\theta) &= -\mathrm{i}U^{-1}(\xi Q + \bar{\xi}\bar{Q})U\Psi(x,\theta) \qquad (4.10)\\
&= -(\xi^\alpha\partial_\alpha + 2\mathrm{i}\theta\sigma^l\bar{\xi}\partial_l)\Psi(x,\theta).
\end{aligned}$$

† $Q, \bar{Q}$ are understood to be operators constructed in terms of field operators A, ψ, F.

4.1.2 Anti-chiral supermultiplet

In this case $D_\alpha\bar{\Phi} = 0$. Proceeding along similar lines we obtain

$$\bar{\Phi}(x,\theta,\bar{\theta}) = U^{-1}\bar{\Psi}(x,\bar{\theta}) = \bar{\Psi}(y^+,\bar{\theta}) = \bar{\Psi}(x^l - \mathrm{i}\theta\sigma^l\bar{\theta},\bar{\theta}) \qquad (4.11)$$

where

$$\bar{\Psi}(x,\bar{\theta}) = \bar{A}(x) + \sqrt{2}\,\bar{\theta}\bar{\psi}(x) + \bar{\theta}^2\bar{F}(x) \qquad (4.12)$$

and

$$\bar{A}(x) = \bar{\Phi}\,| \qquad \sqrt{2}\,\bar{\psi}_{\dot{\alpha}}(x) = \bar{D}_{\dot{\alpha}}\bar{\Phi}\,| \qquad -4\bar{F}(x) = \bar{D}^2\bar{\Phi}\,|. \qquad (4.13)$$

Their SS tranformations may be derived from $\delta\bar{\Phi} = -\mathrm{i}(\xi Q + \bar{\xi}\bar{Q})\bar{\Phi}$ and we find again that $\bar{F}$ transforms into a total divergence. In fact, the 'complex conjugate' of a chiral superfield Φ is an anti-chiral superfield and we will frequently write it simply as $\bar{\Phi}$. We note also that Φ (or $\bar{\Phi}$) necessarily has complex components; a real chiral superfield is a constant (superfield). In fact $\bar{D}_{\dot{\alpha}}\Phi = 0$ implies $D_\alpha\bar{\Phi} = 0$ and if $\Phi = \bar{\Phi}$ (real) it follows that $\{D_\alpha, \bar{D}_{\dot{\alpha}}\}\Phi \sim \partial_l\Phi = 0$. A complex constant chiral superfield satisfies $\bar{D}_{\dot{\alpha}}\Phi = 0$ and $\partial_l\Phi = 0$ and its general form is $c(\theta) = a + \sqrt{2}\,\theta\lambda + \theta^2 f$.

4.1.3 Inverse of chiral superfield

The inverse superfield Φ^{-1} is again chiral and its components are easily derived. Writing

$$\Phi^{-1}(y) = A^{-1}(y) + \sqrt{2}\,\theta\psi^{-1}(y) + \theta^2F^{-1}(y) \qquad (4.14)$$

and using $\Phi^{-1}\Phi = \Phi\Phi^{-1} = 1$ we obtain ($A^{-1} \neq 0$)

$$A^{-1}(y)A(y) = 1$$

$$\begin{aligned} \psi^{-1}(y) &= -A^{-1}\psi(y)A^{-1} = -\frac{1}{A^2}\psi(y) \\ F^{-1}(y) &= -A^{-1}\psi A^{-1}\psi A^{-1} - A^{-1}FA^{-1} = -\frac{1}{A^3}[\psi\psi + AF]. \end{aligned} \qquad (4.15)$$

We proceed analogously for $\bar{\Phi}^{-1}$.

4.2 Supersymmetric action for chiral supermultiplets

4.2.1 Supersymmetric invariants

The construction of an SS invariant action is straightforward if we use superfields. Consider, for example, an interaction term of the form $(\Phi_i\Phi_j\Phi_k\ldots)$. Since this product is a chiral superfield its F-component transforms under SS transformations into a space–time divergence [see (4.8)]. It follows that the integral of the F-component of $(\Phi_i\Phi_j\Phi_k\ldots)$ over the whole of space–time

$$-\tfrac{1}{4}\int d^4x D^2(\Phi_i\Phi_j\Phi_k\ldots)| = -\tfrac{1}{4}\int d^4x\, D^2(\Phi_i\Phi_j\Phi_k\ldots) \qquad (4.16)$$

is an SS invariant. The equality on the RHS arises from the fact that θ-, $\bar{\theta}$-dependent terms in D^2 (and $\bar{D}^2$) are space–time derivatives and we assume that the surface terms do not contribute to the integral. Expressed in terms of component fields, for example,

$$\begin{gathered}\int d^4x(A_iF_j + A_jF_i - \psi_i\psi_j)\\ 3\int d^4x(A^2F - A\psi\psi)\end{gathered} \qquad (4.17)$$

are SS invariants. We note that (4.16) gives rise to SS potential terms involving no space–time derivatives of the fields.

To obtain kinetic energy terms for the matter fields we must also use the anti-chiral complex conjugate superfields $\bar{\Phi}_i$. From a single chiral superfield Φ we may form the following SS invariant (since $\bar{D}^2\bar{\Phi}$ is a chiral superfield):

$$\begin{aligned}\int d^4x D^2[(\bar{D}^2\bar{\Phi})\Phi] &= \int d^4x D^2\bar{D}^2(\bar{\Phi}\Phi)\\ &= \int d^4x D^2\bar{D}^2(\bar{\Phi}\Phi)|\end{aligned} \qquad (4.18)$$

where the last equality follows if we drop the surface terms. Now

$$\begin{aligned}D^2\bar{D}^2(\bar{\Phi}\Phi) &= (D^2\bar{D}^2\bar{\Phi})\Phi + (\bar{D}^2\bar{\Phi})(D^2\Phi) + 2(D^\alpha\bar{D}^2\bar{\Phi})(D_\alpha\Phi)\\ &= ([D^2, \bar{D}^2]\bar{\Phi})\Phi + (\bar{D}^2\bar{\Phi})(D^2\Phi) + 2([D^\alpha, \bar{D}^2]\bar{\Phi})(D_\alpha\Phi).\end{aligned} \qquad (4.19)$$

From (3.25) and using (4.5) after setting $\theta = \bar{\theta} = 0$ we find

$$\begin{aligned}&\tfrac{1}{16}\int d^4x D^2\bar{D}^2(\bar{\Phi}\Phi)\\ &\quad = \int d^4x\{-(\partial_l A)(\partial^l\bar{A}) + \tfrac{1}{2}i[(\partial_l\psi)\sigma^l\bar{\psi} - \psi\sigma^l\partial_l\bar{\psi}] + \bar{F}F\}.\end{aligned} \qquad (4.20)$$

The usual kinetic terms for matter fields are obtained along with a potential term for the auxiliary field which carries canonical dimension 2.

It is convenient for reasons to be explained in Chapter 8 to introduce the

the following notation:

$$-\frac{1}{4}D^2=\int d^2\theta \qquad -\frac{1}{4}\bar{D}^2=\int d^2\bar{\theta}$$
$$\frac{1}{16}D^2\bar{D}^2=\frac{1}{16}\bar{D}^2D^2=\frac{D^\alpha\bar{D}^2D_\alpha}{16}=\frac{\bar{D}_{\dot\alpha}D^2\bar{D}^{\dot\alpha}}{16}=\int d^4\theta \tag{4.21}$$

where an integration over all space–time is suppressed for convenience, and define $d^6s=d^4x\,d^2\theta$, $d^6\bar{s}=d^4x\,d^2\bar{\theta}$, $d^8z=d^4x\,d^4\theta=d^4x\,d^2\theta\,d^2\bar{\theta}$. In terms of superfields the action for the Wess–Zumino model discussed in Chapter 2 then takes the form

$$\int d^8z\bar{\Phi}\Phi+\left(\tfrac{1}{2}m\int d^6s\Phi^2+\tfrac{1}{3}g\int d^6s\Phi^3+\mathrm{CC}\right). \tag{4.22}$$

4.2.2 *General supersymmetric action*

The most general SS action for n chiral superfields Φ_i may be written as

$$I=\int d^8z\bar{\Phi}_i\Phi_i+\int d^6sW(\Phi)+\int d^6\bar{s}\bar{W}(\bar{\Phi}) \tag{4.23}$$

where $W(\Phi)$ is a functional of chiral superfields alone called the *superpotential.* We may easily express (4.23) in terms of component fields. We have

$$\begin{aligned}D^2W(\Phi)\,| &= D^\alpha\left(\frac{\partial W}{\partial\Phi_i}D_\alpha\Phi_i\right)\Bigg|\\ &=\frac{\partial^2 W}{\partial\Phi_j\partial\Phi_i}(D^\alpha\Phi_j)(D_\alpha\Phi_i)\,|+\frac{\partial W}{\partial\Phi_i}D^2\Phi_i\,|\\ &=2\frac{\partial^2W(A)}{\partial A_j\partial A_i}\psi_j^\alpha\psi_{i\alpha}-4\frac{\partial W(A)}{\partial A_i}F_i\end{aligned} \tag{4.24}$$

where $W(A)=W(\Phi_i\to A_i)$ and it depends only on the scalar fields A_i of the supermultiplets (A_i, ψ_i, F_i). From (4.20), (4.23) and (4.24) we find for the SS Lagrangian of n interacting chiral supermultiplets

$$\mathscr{L}=-(\partial_l\bar{A}_i)(\partial^l A_i)+\tfrac{1}{2}\mathrm{i}[(\partial_l\psi_i)\sigma^l\bar{\psi}_i-\psi_i\sigma^l\partial_l\bar{\psi}_i]-V \tag{4.25}$$

where the potential V is given by

$$V=-\left(\bar{F}_iF_i+F_i\frac{\partial W}{\partial A_i}+\bar{F}_i\frac{\partial\bar{W}}{\partial\bar{A}_i}\right)+\frac{1}{2}\left(\frac{\partial^2W}{\partial A_i\partial A_j}\psi_i\psi_j+\mathrm{CC}\right). \tag{4.26}$$

The last term represents a Yukawa interaction with spinor fields while the *scalar potential* V relevant for studying spontaneous SS breaking is

$$V(A_i, \bar{A}_i, F_i, \bar{F}_i) = -\left(\bar{F}_i F_i + F_i \frac{\partial W}{\partial A_i} + \bar{F}_i \frac{\partial \bar{W}}{\partial \bar{A}_i}\right). \tag{4.27}$$

We may eliminate the auxiliary fields F_i using their equations of motion

$$\frac{\partial V}{\partial F_i} = 0 \qquad \frac{\partial V}{\partial \bar{F}_i} = 0$$

or

$$\bar{F}_i = -\frac{\partial W}{\partial A_i} \qquad F_i = -\frac{\partial \bar{W}}{\partial \bar{A}_i}. \tag{4.28}$$

The Lagrangian then takes the form

$$\mathscr{L} = \mathscr{L}_{\mathrm{KE}} + \mathscr{L}_{\mathrm{Yuk}} + \mathscr{L}_{\mathrm{Self}}. \tag{4.29}$$

Here

$$\begin{aligned} \mathscr{L}_{\mathrm{Yuk}} &= -\tfrac{1}{2} M_{ij} \psi_j \psi_i + \mathrm{CC} \\ -\mathscr{L}_{\mathrm{Self}} &= V(A_i, \bar{A}_j) = \bar{F}_i F_i. \end{aligned} \tag{4.30}$$

$\mathscr{L}_{\mathrm{KE}}$ is the kinetic energy term and $M_{ij} = \partial^2 W(A)/\partial A_i \partial A_j$ gives the *fermion mass matrix* when calculated at the minimum of the scalar potential. The *boson mass matrix* is given by

$$M_{\mathrm{B}}^2 = \begin{pmatrix} \dfrac{\partial^2 V}{\partial \bar{A}_i \partial A_j} & \dfrac{\partial^2 V}{\partial \bar{A}_i \partial \bar{A}_j} \\[2ex] \dfrac{\partial^2 V}{\partial A_i \partial A_j} & \dfrac{\partial^2 V}{\partial A_i \partial \bar{A}_j} \end{pmatrix} \tag{4.31}$$

The matrix elements take the form

$$\begin{aligned} \frac{\partial^2 V}{\partial A_i \partial \bar{A}_j} &= \left(\frac{\partial^2 W}{\partial A_i \partial A_k}\right)\left(\frac{\partial^2 \bar{W}}{\partial \bar{A}_j \partial \bar{A}_k}\right) = (M\bar{M})_{ij} \\ \frac{\partial^2 V}{\partial \bar{A}_i \partial A_j} &= \left(\frac{\partial^2 W}{\partial A_j \partial A_k}\right)\left(\frac{\partial^2 \bar{W}}{\partial \bar{A}_i \partial \bar{A}_k}\right) \\ &= (\bar{M}M)_{ij} = (M_{\mathrm{F}}^2)_{ij} \end{aligned} \tag{4.32}$$

$$\frac{\partial^2 V}{\partial A_i \partial A_j} = -\left(\frac{\partial^3 W}{\partial A_i \partial A_j \partial A_k}\right) F_k \qquad \frac{\partial^2 V}{\partial \bar{A}_i \partial \bar{A}_j} = -\left(\frac{\partial^3 W}{\partial \bar{A}_i \partial \bar{A}_j \partial \bar{A}_k}\right) \bar{F}_k.$$

We find

$$\mathrm{Tr} M_{\mathrm{B}}^2 = 2\mathrm{Tr}(\bar{M}M) = 2\mathrm{Tr} M_{\mathrm{F}}^2. \tag{4.33}$$

For the supersymmetric ground state $V = 0$ implies that $F_i = \bar{F}_i = 0$ and bosons and fermions have the same mass†. In the case of a spontaneous breakdown of SS (see Chapter 6) we must diagonalise the mass matrices and we find different masses for the component fields of a supermultiplet.

A renormalisable action is obtained for

$$W(\Phi) = \lambda_i \Phi_i + \tfrac{1}{2} m_{ij} \Phi_i \Phi_j + \tfrac{1}{3} g_{ijk} \Phi_i \Phi_j \Phi_k \tag{4.34}$$

where m_{ij} and g_{ijk} are completely symmetric tensors. From

$$W(A) = \lambda_i A_i + \tfrac{1}{2} m_{ij} A_i A_j + \tfrac{1}{3} g_{ijk} A_i A_j A_k \tag{4.35}$$

we find

$$\begin{aligned} -\bar{F}_i &= \lambda_i + m_{ij} A_j + g_{ijk} A_j A_k \\ M_{ij} &= m_{ij} + 2 g_{ijk} A_k \\ \frac{\partial^3 W}{\partial A_i \partial A_j \partial A_k} &= 2 g_{ijk}. \end{aligned} \tag{4.36}$$

For the Wess–Zumino model $W(A) = \frac{1}{2} m A^2 + \frac{1}{3} g A^3$, $\bar{F} = -(mA + gA^2)$, $M = (m + 2gA)$ and $M\bar{M} = \bar{M}M = |m + 2gA|^2$. In the SS ground states, $\bar{F} = 0$, we verify that the boson and fermion fields have the same mass.

4.2.3 *R or chiral transformation, R-invariance*

The SS algebra remains unaltered under the chiral phase or R-transformation of the Majorana spinor generators $Q_\alpha \to e^{-i\alpha} Q_\alpha$, $\bar{Q}_{\dot{\alpha}} \to e^{i\alpha} \bar{Q}_{\dot{\alpha}}$ (§1.3). The generator R corresponding to this symmetry does not commute with SS generators, $[R, Q_\alpha] = -Q_\alpha$, $[R, \bar{Q}_{\dot{\alpha}}] = \bar{Q}_{\dot{\alpha}}$, $[R, P_l] = [R, M_{lm}] = 0$. On the superspace it may be represented by the differential operator $R = (\theta\partial - \bar{\theta}\bar{\partial})$ and R-transformation corresponds to the chiral phase transformation on the Majorana spinor coordinates $\theta \to e^{i\alpha}\theta = \theta'$, $\bar{\theta} \to e^{-i\alpha}\bar{\theta} = \bar{\theta}'$, $\delta\theta = i\alpha\theta = i\alpha R\theta$, $\delta\bar{\theta} = -i\alpha\bar{\theta} = i\alpha R\bar{\theta}$. We may also obtain a realisation of R on scalar superfields by defining a chiral or R transformation on them as follows:

$$\begin{aligned} S'(x, \theta', \bar{\theta}') &= e^{i2n\alpha} S(x, \theta, \bar{\theta}) = e^{i2n\alpha} S(x, e^{-i\alpha}\theta', e^{i\alpha}\bar{\theta}') \\ \bar{S}'(x, \theta', \bar{\theta}') &= e^{-i2n\alpha} \bar{S}(x, \theta, \bar{\theta}) \end{aligned} \tag{4.37}$$

where $2n$ is the chiral or R charge of the superfield S while it is $(-2n)$ for the complex conjugate $\bar{S}$. We derive for the infinitesimal transformations

$$\delta_R S = -i\alpha(R - 2n) S(x, \theta, \bar{\theta}) \qquad \delta_R \bar{S} = -i\alpha(R + 2n) \bar{S}(x, \theta, \bar{\theta}) \tag{4.38}$$

† This is also clear from $[P_l P^l, Q_\alpha] = [P_l P^l, \bar{Q}_{\dot{\alpha}}] = 0$.

where $R = (\theta\partial - \bar{\theta}\bar{\partial})$. For a real superfield we find $n = 0$. From $[R, \mathrm{D}_\alpha] = -\mathrm{D}_\alpha$, $[R, \bar{\mathrm{D}}_{\dot{\alpha}}] = \bar{\mathrm{D}}_{\dot{\alpha}}$ it is clear that under R-transformation a chiral (anti-chiral) superfield stays chiral (anti-chiral). The chiral transformations of the component fields may be derived as usual.

Consider, for example, the chiral superfield, $\Phi = U\Psi(x, \theta)$; we obtain

$$\delta_{\mathrm{R}}\Phi = U\delta_{\mathrm{R}}\Psi(x, \theta) = -\mathrm{i}\alpha(R - 2n)U\Psi(x, \theta)$$

or

$$\begin{aligned}\delta_{\mathrm{R}}\Psi(x, \theta) &= -\mathrm{i}\alpha U^{-1}(R - 2n)U\Psi(x, \theta)\\ &= -\mathrm{i}\alpha(R - 2n)\Psi(x, \theta)\end{aligned} \tag{4.39}$$

a result which could also be derived directly from $\Psi'(x, \theta') = \mathrm{e}^{\mathrm{i}2n\alpha}\Psi(x, \theta)$. The last line follows from the identity $U^{-1}(\theta\partial - \bar{\theta}\bar{\partial})U = (\theta\partial - \bar{\theta}\bar{\partial})$. A straightforward calculation gives

$$\delta_{\mathrm{R}}\Psi(x, \theta) = -\mathrm{i}\alpha[-2n\Psi(x, \theta) + \sqrt{2}\,\theta\psi + 2\theta^2 F]. \tag{4.40}$$

For an anti-chiral superfield we obtain

$$\delta_{\mathrm{R}}\bar{\Psi}(x, \bar{\theta}) = -\mathrm{i}\alpha[2n\bar{\Psi} - \sqrt{2}\,\bar{\theta}\bar{\psi} - 2\bar{\theta}^2\bar{F}]. \tag{4.41}$$

Hence for the component fields the chiral transformations are (see(2.40))

$$\begin{aligned}(A, \psi, F) &\to \mathrm{e}^{\mathrm{i}2n\alpha}(A, \mathrm{e}^{-\mathrm{i}\alpha}\psi, \mathrm{e}^{-\mathrm{i}2\alpha}F)\\ (\bar{A}, \bar{\Psi}, \bar{F}) &\to \mathrm{e}^{-\mathrm{i}2n\alpha}(\bar{A}, \mathrm{e}^{\mathrm{i}\alpha}\bar{\psi}, \mathrm{e}^{\mathrm{i}2\alpha}\bar{F}).\end{aligned} \tag{4.42}$$

The bosonic and fermionic components transform differently under an R transformation which consequently cannot commute with an SS transformation. The chiral charges of the component fields may, alternatively, be derived by simply giving chiral charges to $\theta, \bar{\theta}$, viz, $R(\theta) = 1, R(\bar{\theta}) = -1$. Requiring that the total chiral charge of each term in the expansion of the superfield be equal to the chiral charge of the superfield the results (4.42) immediately follow. It is very simple to check the R-invariance of a theory written in terms of superfields if we assign $R(\mathrm{d}\theta) = -1$, $R(d\bar{\theta}) = +1, R[\mathrm{D}_\alpha] = R[\partial_\alpha] = -1, R[\bar{\mathrm{D}}_{\dot{\alpha}}] = R[\bar{\partial}_{\dot{\alpha}}] = 1$. The kinetic term $\int \mathrm{d}^4\theta\bar{\Phi}\Phi$ is clearly R-invariant; it may be verified in component formulation as well.

Requiring R-invariance of the action puts restrictions on the form of the interaction terms. For example, in the Wess–Zumino model the Φ^2 term will be absent if $n \neq \frac{1}{2}$ and a Φ^3 term is allowed only if $6n - 2 = 0$ or $n = \frac{1}{3}$. Many interesting SS theories are found to be R-invariant and thus carry an extra $U(1)_{\mathrm{R}}$ symmetry. The identification of an R-symmetry may lead to the identification in the theory of a large symmetry group. We will discuss more about R-invariance in Chapter 5.

4.3 The supercurrent superfield

The supercurrent multiplet for the Wess–Zumino action was discussed in component formulation in §2.2.2. We now describe the corresponding superfield formulation. We may use the definitions (4.5) to recast the axial current (2.41) in terms of the chiral superfield Φ

$$J_5^n = -\tfrac{1}{2}\bar{\sigma}^{n\dot{\alpha}\beta}(\tfrac{1}{3}V_{\beta\dot{\alpha}})|_{\theta=\bar{\theta}=0} \tag{4.43}$$

where $V_{\alpha\dot{\alpha}}(z)$ is a vector superfield ($q=-\tfrac{1}{3}$)

$$V_{\beta\dot{\alpha}} = (\bar{D}_{\dot{\alpha}}\bar{\Phi})(D_\beta\Phi) - 2i(\Phi\partial_{\beta\dot{\alpha}}\bar{\Phi} - \bar{\Phi}\partial_{\beta\dot{\alpha}}\Phi) \tag{4.44}$$

which is a candidate for the supercurrent superfield. In fact we find

$$\begin{aligned}\bar{D}^{\dot{\alpha}}V_{\beta\dot{\alpha}} &= -(\bar{D}^2\bar{\Phi})(D_\beta\Phi) - (\bar{D}_{\dot{\alpha}}\bar{\Phi})(\bar{D}^{\dot{\alpha}}D_\beta\Phi) - 2i\Phi\overleftrightarrow{\partial}_{\beta\dot{\alpha}}\bar{D}^{\dot{\alpha}}\bar{\Phi}\\ &= -(\bar{D}^2\bar{\Phi})(D_\beta\Phi) + \tfrac{1}{2}\Phi D_\beta\bar{D}^2\bar{\Phi}\end{aligned} \tag{4.45}$$

where we use $\bar{D}_{\dot{\alpha}}D_\beta\Phi = \{\bar{D}_{\dot{\alpha}}, D_\beta\}\Phi = -2i\partial_{\beta\dot{\alpha}}\Phi$ and $\tfrac{1}{4}[D_\alpha, \bar{D}^2] = -i\partial_{\alpha\dot{\beta}}\bar{D}^{\dot{\beta}}$ as discussed in (3.25). We may also easily derive the equations of motion, $\delta I/\delta S = 0$, from (4.22) if we rewrite the action in terms of the unconstrained superfields S, $\bar{S}$ defined by $\Phi = -\tfrac{1}{4}\bar{D}^2S$, $\bar{\Phi} = -\tfrac{1}{4}D^2\bar{S}$.
We find†

$$\tfrac{1}{4}\bar{D}^2\bar{\Phi} = m\Phi + g\Phi^2 \tag{4.46}$$

and (4.45) then results in

$$\bar{D}^{\dot{\alpha}}V_{\beta\dot{\alpha}} = (D_\beta U) \qquad \bar{D}_{\dot{\alpha}}U = 0 \qquad U = -m\Phi^2. \tag{4.47}$$

We recall that $m\Phi^2$ is the 'chiral multiplet of anomalies' (2.55) discussed in §2.2.2. It is straightforward to derive the component currents contained in the Poincaré supercurrent superfield V and their SS variations. For the massless case when $U=0$ we obtain the conformal current superfield.

Problems

4.1 Show that

$$\begin{aligned}\Phi &= A(x) + \sqrt{2}\,\theta\psi(x) + \theta^2F(x) + i\theta\sigma^l\bar{\theta}\partial_lA(x)\\ &\quad + \frac{i}{\sqrt{2}}\theta^2\bar{\theta}\bar{\sigma}^l\partial_l\psi(x) + \tfrac{1}{4}\theta^2\bar{\theta}^2\Box A(x)\end{aligned}$$

$$\begin{aligned}\bar{\Phi} &= \bar{A}(x) + \sqrt{2}\,\bar{\theta}\bar{\psi}(x) + \bar{\theta}^2\bar{F}(x) - i\theta\sigma^l\bar{\theta}\partial_l\bar{A}(x)\\ &\quad + \frac{i}{\sqrt{2}}\bar{\theta}^2\theta\sigma^l\partial_l\bar{\psi}(x) + \tfrac{1}{4}\bar{\theta}^2\theta^2\Box\bar{A}(x).\end{aligned}$$

† See also Chapter 9.

4.2 Show that (4.9) implies the x-independence of the Green functions $\langle TA_1(x_1)A_2(x_2)\ldots A_m(x_m)\rangle_0$ where $|0\rangle$ is a supersymmetric ground state and A_i are assumed to commute at equal times.

4.3 Show that the SS and R-transformations may be realised on the superfields $\Psi(x,\theta)$ defined over the (left) chiral superspace spanned by x and θ. Hints: See(4.10) and problem 3.6.

4.4 The differential constraints defining the chiral superfields may be solved by introducing unconstrained superpotential superfields S, $\bar{S}$ such that $\Phi_i = -\frac{1}{4}\bar{D}^2 S_i$, $\bar{\Phi}_i = -\frac{1}{4}D^2\bar{S}_i$. Write the action (4.23) in terms of S_i, $\bar{S}_i$ and derive the equations of motion for W given in (4.34).

4.5 Find the supersymmetric ground states of the Wess–Zumino model (4.22) by minimising the scalar potential. Verify that the two real scalar fields and the fermionic field all carry the same mass.

4.6 Compute the SS variation of $D^2(\Phi_i\Phi_j)$ and $D^2\bar{D}^2(\bar{\Phi}\Phi)$.

4.7 Rewrite (4.29) for W given in (4.34) in four-component notation and in terms of the real scalar fields $\tilde{A}_i, \tilde{B}_i, \tilde{F}_i, \tilde{G}_i$ defined by $A_i = (1/\sqrt{2})(\tilde{A}_i - \mathrm{i}\tilde{B}_i)$ and $F_i = (1/\sqrt{2})(\tilde{F}_i + \mathrm{i}\tilde{G}_i)$. (For one chiral superfield it reads as

$$\mathscr{L} = -\frac{1}{2}(\partial_l\tilde{A})(\partial^l\tilde{A}) - \frac{1}{2}(\partial_l\tilde{B})(\partial^l\tilde{B}) + \frac{\mathrm{i}}{2}\bar{\psi}(\gamma^l\partial_l + m)\psi + \frac{1}{2}(\tilde{F}^2 + \tilde{G}^2)$$

$$+ m(\tilde{A}\tilde{F} + \tilde{B}\tilde{G}) + \frac{g}{\sqrt{2}}[(\tilde{A}^2 - \tilde{B}^2)\tilde{F} + 2\tilde{G}\tilde{A}\tilde{B} + \mathrm{i}\bar{\psi}(\tilde{A} - \tilde{B}\gamma_5)\psi]$$

while the SS transformations are

$$\delta\tilde{A} = \mathrm{i}\bar{\xi}\psi \qquad \delta\tilde{B} = \mathrm{i}\bar{\xi}\gamma_5\psi \qquad \delta\tilde{F} = \mathrm{i}\bar{\xi}\gamma^l\partial_l\psi \qquad \delta\tilde{G} = \mathrm{i}\bar{\xi}\gamma_5\gamma^l\partial_l\psi$$

$$\delta\psi = -[(\tilde{F} + \gamma_5\tilde{G}) + \gamma^l\partial_l(\tilde{A} + \gamma_5\tilde{B})]\xi$$

where we use $\delta\bar{\psi}^{\dot{\alpha}} = -\sqrt{2}[\bar{\xi}^{\dot{\alpha}}\bar{F} + \mathrm{i}(\bar{\sigma}^l\xi)^{\dot{\alpha}}\partial_l\bar{A}]$).

4.8 Show that the components of an anti-chiral supermultiplet have the SS transformation law given by

$$\delta\bar{A} = -\sqrt{2}\,\bar{\xi}\bar{\psi} \qquad \delta\bar{\psi}^{\dot{\alpha}} = -\sqrt{2}[\bar{\xi}^{\dot{\alpha}}\bar{F} + \mathrm{i}(\bar{\sigma}^l\xi)^{\dot{\alpha}}\partial_l\bar{A}]$$

$$\delta\bar{F} = -\sqrt{2}\,\xi^{\alpha}(\mathrm{i}\sigma^l\partial_l\bar{\psi})_{\alpha}$$

while the *kinetic supermultiplet* defined by

$$T\Phi \equiv (\bar{F}, \mathrm{i}(\sigma^l\partial_l\bar{\psi})_{\alpha}, \Box\bar{A}) \qquad TT\Phi = \Box\Phi$$

transforms as a chiral supermultiplet.

4.9 Compute the product $\Phi(T\Phi)$ to show that the kinetic energy term (4.20) is the F-component of this product. Compute $\bar{D}^2\bar{\Phi}$ also.

5

GAUGE SUPERFIELD—ABELIAN CASE

5.1 Vector or gauge supermultiplet

A scalar superfield satisfying the reality constraint $V(z) = \overline{V(z)}$ gives rise to an irreducible supermultiplet which contains in it a vector gauge field $v_l(x)$. The most general expression of the vector superfield is

$$\begin{aligned} V(x, \theta, \bar{\theta}) = {} & C(x) + \mathrm{i}\theta\chi(x) - \mathrm{i}\bar{\theta}\bar{\chi}(x) + \frac{\mathrm{i}}{2}\theta^2(M(x) + \mathrm{i}N(x)) \\ & - \frac{\mathrm{i}}{2}\bar{\theta}^2(M(x) - \mathrm{i}N(x)) - \theta\sigma^l\bar{\theta}v_l(x) \\ & + \mathrm{i}\theta^2\bar{\theta}\left(\bar{\lambda}(x) + \frac{\mathrm{i}}{2}\bar{\sigma}^l\partial_l\chi(x)\right) \\ & - \mathrm{i}\bar{\theta}^2\theta\left(\lambda(x) + \frac{\mathrm{i}}{2}\sigma^l\partial_l\bar{\chi}(x)\right) + \frac{1}{2}\theta^2\bar{\theta}^2\left(D(x) + \frac{1}{2}\Box\, C(x)\right) \end{aligned} \tag{5.1}$$

where† C, D, M, N and v_l are real. From the unit canonical dimension of v_l we deduce $[V(z)] = 0$. Then $[C] = 0$, $[\chi] = [\bar{\chi}] = \frac{1}{2}$, $[\lambda] = [\bar{\lambda}] = \frac{3}{2}$, $[M] = [N] = 1$ and $[D] = 2$. A vector superfield may be constructed, for example, from a chiral superfield Λ and its complex conjugate $\bar{\Lambda}$. Writing

$$\begin{aligned} \Lambda(x, \theta, \bar{\theta}) &= U[\tilde{A}(x) + \sqrt{2}\,\theta\tilde{\psi}(x) + \theta^2\tilde{F}(x)] \\ \bar{\Lambda}(x, \theta, \bar{\theta}) &= U^{-1}[\bar{\tilde{A}}(x) + \sqrt{2}\,\bar{\theta}\bar{\tilde{\psi}}(x) + \bar{\theta}^2\bar{\tilde{F}}(x)] \\ \bar{\mathrm{D}}_{\dot{\alpha}}\Lambda &= \mathrm{D}_\alpha\bar{\Lambda} = 0 \end{aligned} \tag{5.2}$$

then

$$\begin{aligned} \mathrm{i}(\Lambda - \bar{\Lambda}) = {} & \mathrm{i}(\tilde{A} - \bar{\tilde{A}}) + \mathrm{i}\sqrt{2}\,(\theta\tilde{\psi} - \bar{\theta}\bar{\tilde{\psi}}) + \mathrm{i}\theta^2\tilde{F} - \mathrm{i}\bar{\theta}^2\bar{\tilde{F}} \\ & - \theta\sigma^l\partial_l(\tilde{A} + \bar{\tilde{A}}) - \frac{1}{\sqrt{2}}\theta^2\bar{\theta}\bar{\sigma}^l\partial_l\tilde{\psi} + \frac{1}{\sqrt{2}}\bar{\theta}^2\theta\sigma^l\partial_l\bar{\tilde{\psi}} \\ & + \frac{\mathrm{i}}{4}\theta^2\bar{\theta}^2\,\Box\,(\tilde{A} - \bar{\tilde{A}}) \end{aligned} \tag{5.3}$$

† Note that $\theta\sigma^l\bar{\theta} = (\mathrm{i}/2)\bar{\theta}\gamma_5\gamma^l\theta$ and consequently if v_m is a polar vector the fields C, D and N are pseudoscalars while M is a scalar.

is a vector superfield. Similar to the case of a chiral superfield, the reality constraint does not define $V(x, \theta, \bar{\theta})$ uniquely and we may define a supersymmetric generalisation of a gauge transformation (Abelian case) by

$$V \to V + \frac{\mathrm{i}}{g_1}(\Lambda - \bar{\Lambda}) \tag{5.4}$$

where g_1 is the gauge coupling constant†. The corresponding transformation of the vector field v_l, the two Majorana spinors λ, χ and the three scalar fields D, C, M, N are given by ($g_1 = 1$)

$$C \to C + \mathrm{i}(\tilde{A} - \bar{\tilde{A}}) \qquad \chi \to \chi + \sqrt{2}\,\tilde{\psi}$$

$$M \to M + (\tilde{F} + \bar{\tilde{F}}) \qquad N \to N - \mathrm{i}(\tilde{F} - \bar{\tilde{F}})$$

$$D \to D \qquad \lambda \to \lambda \qquad v_l \to v_l + \partial_l(\tilde{A} + \bar{\tilde{A}})$$

$$F_{lm} \to F_{lm} \equiv (\partial_l v_m - \partial_m v_l). \tag{5.5}$$

It contains the usual Abelian gauge transformation of the vector field v_l while D, λ, F_{lm} are invariant under the SS generalisation of gauge transformations.

5.2 Supersymmetry transformations of component fields

The component fields may be expressed as covariant derivatives of V evaluated at $\theta = \bar{\theta} = 0$. We find

$$V| = C \qquad \mathrm{D}_\alpha V| = \mathrm{i}\chi_\alpha \qquad \bar{\mathrm{D}}_{\dot{\alpha}} V = -\mathrm{i}\bar{\chi}_{\dot{\alpha}}$$

$$\mathrm{D}^2 V| = -2\mathrm{i}(M + \mathrm{i}N) \qquad \bar{\mathrm{D}}^2 V| = 2\mathrm{i}(M - \mathrm{i}N)$$

$$\bar{\mathrm{D}}^2 \mathrm{D}_\alpha V| = -\bar{\partial}^2 \mathrm{D}_\alpha V| = 4\mathrm{i}\lambda_\alpha$$

$$\mathrm{D}^2 \bar{\mathrm{D}}_{\dot{\alpha}} V| = -\partial^2 \bar{\mathrm{D}}_{\dot{\alpha}} V| = -4\mathrm{i}\bar{\lambda}_{\dot{\alpha}}$$

$$\mathrm{D}^\alpha \bar{\mathrm{D}}^2 \mathrm{D}_\alpha V| = \partial^\alpha \bar{\mathrm{D}}^2 \mathrm{D}_\alpha V| = 8D$$

$$[\mathrm{D}_\alpha, \bar{\mathrm{D}}_{\dot{\alpha}}]\, V| = 2\partial_\alpha \bar{\partial}_{\dot{\alpha}} V| = -2\sigma^l_{\alpha\dot{\alpha}} v_l$$

$$\mathrm{D}^\beta \bar{\mathrm{D}}^2 \mathrm{D}_\alpha V| = \partial^\beta \bar{\mathrm{D}}^2 \mathrm{D}_\alpha V| = 4\delta^\beta_\alpha D - 2\mathrm{i}(\sigma^m \bar{\sigma}^l)_\alpha{}^\beta F_{ml} \tag{5.6}$$

where $F_{lm} = (\partial_l v_m - \partial_m v_l)$. The SS transformations may then be derived

† Note that $[\Lambda] = [\bar{\Lambda}] = 0$ so that Λ, $\bar{\Lambda}$ are gauge parameter superfields rather than matter superfields.

straightforwardly:

$$\begin{aligned}
\delta C &= \delta V| = -\mathrm{i}(\xi Q + \bar{\xi}\bar{Q})V| = -(\xi D + \bar{\xi}\bar{D})V| = -\mathrm{i}(\xi\chi - \bar{\xi}\bar{\chi}) \\
\delta\chi_\alpha &= \mathrm{i}(\xi D + \bar{\xi}\bar{D})\mathrm{D}_\alpha V| = -\xi_\alpha(M + \mathrm{i}N) - \mathrm{i}(\sigma^l\bar{\xi})_\alpha(v_l - \mathrm{i}\partial_l C) \\
\delta\bar{\chi}_{\dot\alpha} &= -\bar{\xi}_{\dot\alpha}(M - \mathrm{i}N) + \mathrm{i}(\xi\sigma^l)_{\dot\alpha}(v_l + \mathrm{i}\partial_l C) \\
\delta(M + \mathrm{i}N) &= \frac{\mathrm{i}}{2}\,\delta\mathrm{D}^2 V| = -\frac{\mathrm{i}}{2}(\xi D + \bar{\xi}\bar{D})\mathrm{D}^2 V| = \frac{\mathrm{i}}{2}\,\bar{\xi}^{\dot\alpha}\bar{\mathrm{D}}_{\dot\alpha}\mathrm{D}^2 V| \\
&= \frac{\mathrm{i}}{2}\,\bar{\xi}^{\dot\alpha}([\bar{\mathrm{D}}_{\dot\alpha}, \mathrm{D}^2] + \mathrm{D}^2\bar{\mathrm{D}}_{\dot\alpha})V| = \frac{\mathrm{i}}{2}\,\bar{\xi}^{\dot\alpha}(4\mathrm{i}\sigma^l_{\alpha\dot\alpha}\mathrm{D}^\alpha\partial_l + \mathrm{D}^2\bar{\mathrm{D}}_{\dot\alpha})V| \\
&= 2\bar{\xi}\bar{\lambda} + 2\mathrm{i}\partial_l(\chi\sigma^l\bar{\xi}) \neq \text{total divergence} \\
\delta\lambda_\alpha &= -\frac{1}{4\mathrm{i}}(\xi^\beta\mathrm{D}_\beta)\bar{\mathrm{D}}^2\mathrm{D}_\alpha V| = \mathrm{i}\xi_\alpha D - \frac{1}{2}(\sigma^m\bar{\sigma}^l)_\alpha{}^\beta\xi_\beta F_{ml} \\
\delta v_l &= -\mathrm{i}(\xi\sigma_l\bar{\lambda} - \lambda\sigma_l\bar{\xi}) - \partial_l(\xi\chi + \bar{\xi}\bar{\lambda}) \\
\delta F_{lm} &= -\mathrm{i}[\partial_l(\xi\sigma_m\bar{\lambda} - \lambda\sigma_m\bar{\xi}) - \partial_m(\xi\sigma_l\bar{\lambda} - \lambda\sigma_l\bar{\xi})] \\
\delta D &= \partial_l(\xi\sigma^l\bar{\lambda} - \bar{\xi}\bar{\sigma}^l\lambda).
\end{aligned} \tag{5.7}$$

The component D with the highest dimension again transforms into a total divergence. The fields λ_α, F_{lm} and D are invariant under supergauge transformations (5.5) while under SS transformations they transform among themselves and constitute an irreducible supermultiplet.

5.3 Field strength supermultiplet

From the observation in the previous section it is clear that the supermultiplet $(\lambda_\alpha, \bar{\lambda}_{\dot\alpha}, F_{lm}, D)$ is a candidate for the field strength supermultiplet. The Majorana spinor field λ_α carries the lowest dimension $[\lambda_\alpha] = \frac{3}{2}$ while $[D] = [F_{lm}] = 2$. Hence it suggests that a spinor superfield W_α contains the supermultiplet in question. We should also require that W_α be invariant under supergauge transformations (5.4). Making use of covariant derivatives† which carry dimension $[\mathrm{D}_\alpha] = [\bar{\mathrm{D}}_{\dot\alpha}] = \frac{1}{2}$ a suitable candidate is found to be a chiral spinor superfield

$$\begin{aligned}
&W_\alpha = \bar{\mathrm{D}}^2\mathrm{D}_\alpha V \qquad &&\bar{\mathrm{D}}_{\dot\beta}W_\alpha = 0 \qquad &&[W_\alpha] = \tfrac{3}{2} \\
&\bar{W}_{\dot\alpha} = \mathrm{D}^2\bar{\mathrm{D}}_{\dot\alpha}V \qquad &&\mathrm{D}_\beta\bar{W}_{\dot\alpha} = 0 \qquad &&[\bar{W}_{\dot\alpha}] = \tfrac{3}{2}.
\end{aligned} \tag{5.8}$$

† Covariant derivatives map superfield to superfield.

This expression is also suggested by (5.6) where $\lambda_\alpha \cong \bar{D}^2 D_\alpha V|$. The invariance under (5.4) follows from

$$W_\alpha \to W_\alpha + i\bar{D}^2 D_\alpha \Lambda = W_\alpha + i[\bar{D}^2, D_\alpha]\Lambda = W_\alpha \tag{5.9}$$

since $[\bar{D}^2, D_\alpha] = 4i\partial_{\alpha\dot{\alpha}}\bar{D}^{\dot{\alpha}}$ and $\bar{D}_{\dot{\alpha}}\Lambda = 0$.

The component fields contained in the chiral superfield W_α may be readily obtained from (4.5) and (5.6):

$$\begin{aligned} W_\alpha| &= 4i\lambda_\alpha(x) \\ D^\beta W_\alpha| &= D^\beta \bar{D}^2 D_\alpha V| = +4\,\delta^\beta_\alpha D - 2i(\sigma^l\bar{\sigma}^m)_\alpha{}^\beta F_{lm} \\ -\tfrac{1}{4}D^2 W_\alpha| &= -\tfrac{1}{4}D^2\bar{D}^2 D_\alpha V| = -\tfrac{1}{4}D^2[\bar{D}^2, D_\alpha]V| \\ &= -i\sigma^l_{\alpha\dot{\alpha}}\partial_l D^2\bar{D}^{\dot{\alpha}}V| = -4\partial_{\alpha\dot{\alpha}}\bar{\lambda}^{\dot{\alpha}}. \end{aligned} \tag{5.10}$$

Consequently

$$W_\alpha(z) = U[W_\alpha| + \theta^\beta(D_\beta W_\alpha|) + \theta^2(-\tfrac{1}{4}D^2 W_\alpha|)] \tag{5.11}$$

or

$$\begin{aligned} -\tfrac{1}{4}W_\alpha(x,\theta,\bar{\theta}) = -i\lambda_\alpha(y) + \theta_\beta\left(\delta^\beta_\alpha D(y) + \frac{i}{2}(\sigma^l\bar{\sigma}^m)_\alpha{}^\beta F_{lm}(y)\right) \\ + \theta^2\partial_{\alpha\dot{\alpha}}\bar{\lambda}^{\dot{\alpha}}(y). \end{aligned} \tag{5.12}$$

Alternatively the result could be obtained on using

$$U^{-1}W_\alpha = U^{-1}\bar{D}^2 U U^{-1} D_\alpha V = -\bar{\partial}^2(U^{-1}D_\alpha V) \tag{5.13}$$

and calculating the $\bar{\theta}^2$ component of $U^{-1}D_\alpha V$. By straightforward tedious computation

$$\begin{aligned} D_\alpha V &= -(\sigma^l\bar{\theta})_\alpha v_l + 2i\theta_\alpha\bar{\theta}\bar{\lambda} - i\bar{\theta}^2\lambda_\alpha + \theta_\alpha\bar{\theta}^2 D \\ &\quad - \frac{i}{2}\partial_{\alpha\dot{\beta}}[-\theta^\beta\sigma^m{}_\beta{}^{\dot{\beta}}v_m + i\theta^2\bar{\lambda}^{\dot{\beta}}]\bar{\theta}^2 \\ U^{-1}D_\alpha V &= \bar{\theta}^2(\ldots) + \text{terms linear in } \bar{\theta}_{\dot{\alpha}} \end{aligned} \tag{5.14}$$

where in the last line the parenthesis contains the expression on the RHS of (5.12) calculated at the coordinate x^l. Hence we obtain (5.12).

From (5.8) and (3.25) we obtain

$$D^\alpha W_\alpha = \bar{D}_{\dot{\alpha}}\bar{W}^{\dot{\alpha}} = -8\,\Box\, P_T V \tag{5.15}$$

representing a transverse vector multiplet as discussed in §3.4

We also note that V^2 is not invariant under generalised gauge transformations (5.4) and its $\theta^2\bar{\theta}^2$ component is given by

$$\begin{aligned} V^2|_{\theta^2\bar{\theta}^2} = -\frac{1}{2}v^l v_l - \chi\lambda - \bar{\chi}\bar{\lambda} + \frac{1}{2}(M^2+N^2) - \frac{i}{2}\chi\sigma^l\partial_l\bar{\chi} \\ -\frac{i}{2}\bar{\chi}\bar{\sigma}^l\partial_l\chi + \frac{1}{2}C\Box C + CD. \end{aligned} \tag{5.16}$$

From the discussion to follow, this term gives rise to equal masses for the two spin-$\frac{1}{2}$ fields, the gauge field v_l and the scalar field C. From the considerations of dimensions it is clear that C and χ are gauge fields and M, N, D appear as auxiliary fields. When the V^2 term is not present the 'physical' fields are v_l (photon) and the spin-$\frac{1}{2}$ field $(\lambda_\alpha, \bar{\lambda}_{\dot{\alpha}})$ which describes a photino.

5.4 Kinetic energy term for vector supermultiplet

The kinetic term invariant under SS transformations and the SS generalisation of gauge transformations (5.4) may be constructed as follows.

From the fact that W_α is a chiral superfield we know that $\mathrm{D}^2 W^\alpha W_\alpha|$ transforms under SS transformations into a total space–time derivative. It is hence a candidate for the kinetic term in the Lagrangian. From

$$\mathrm{D}^2 W^\alpha W_\alpha = 2[(\mathrm{D}^2 W^\alpha)W_\alpha - (\mathrm{D}^\beta W^\alpha)(\mathrm{D}_\beta W_\alpha)] \tag{5.17}$$

and (5.10) we find

$$-\tfrac{1}{4}\mathrm{D}^2 W^\alpha W_\alpha| = 32[-\tfrac{1}{4}F^{lm}F_{lm} - \mathrm{i}\lambda\sigma^l\partial_l\bar{\lambda} + \tfrac{1}{2}D^2 - \tfrac{1}{4}{}^*F^{lm}F_{lm}]. \tag{5.18}$$

We may thus write the required kinetic term following the discussion and notation introduced in Chapter 4 as

$$\begin{aligned} I_V &= \frac{1}{32}\int \mathrm{d}^4x(-\tfrac{1}{4}\mathrm{D}^2)W^\alpha W_\alpha = \frac{1}{32}\int \mathrm{d}^4x\,\mathrm{d}^2\theta\, W^\alpha W_\alpha \\ &= \int \mathrm{d}^4x(-\tfrac{1}{4}F^{lm}F_{lm} - \mathrm{i}\lambda\sigma^l\partial_l\bar{\lambda} + \tfrac{1}{2}D^2) \end{aligned} \tag{5.19}$$

where we drop the surface term involving the dual ${}^*F^{lm} = (\mathrm{i}/2)\varepsilon^{lmpq}F_{pq}$, ${}^*({}^*F^{lm}) = F^{lm}$. Alternatively, we may add the CC term $\sim\bar{\mathrm{D}}^2\bar{W}_{\dot{\alpha}}\bar{W}^{\dot{\alpha}}$. The usual expressions of KE terms for photon and photino fields are present along with a term for the auxiliary field D of dimension 2. From (3.25) and (5.15) it follows that

$$\begin{aligned} \mathrm{D}^2 W^\alpha W_\alpha &= \mathrm{D}^2\bar{\mathrm{D}}^2\{(\mathrm{D}^\alpha V)(\bar{\mathrm{D}}^2\mathrm{D}_\alpha V)\} = \mathrm{D}^2\bar{\mathrm{D}}^2\{\mathrm{D}^\alpha(VW_\alpha) - V\mathrm{D}^\alpha W_\alpha\} \\ &= [\mathrm{D}^2, \bar{\mathrm{D}}^2]\mathrm{D}^\alpha(VW_\alpha) - \mathrm{D}^2\bar{\mathrm{D}}^2(V\mathrm{D}^\alpha W_\alpha) \\ &= 8\mathrm{D}^2\bar{\mathrm{D}}^2(V\Box P_{\mathrm{T}}V) + \text{surface terms.} \end{aligned} \tag{5.20}$$

On dropping the surface terms we may rewrite

$$\begin{aligned} I_V &= -\int \mathrm{d}^4x\left(\frac{\mathrm{D}^2\bar{\mathrm{D}}^2}{16}\right)V\Box P_{\mathrm{T}}V = -\int \mathrm{d}^8z\, V\Box P_{\mathrm{T}}V \\ &= \frac{1}{8}\int \mathrm{d}^8z\, V\mathrm{D}^\alpha W_\alpha. \end{aligned} \tag{5.21}$$

5.5 Fayet–Illiopoulos† and mass terms

From (5.7) it follows that the space–time integral of the D-term of a vector superfield is an SS invariant. For the (Abelian) case under discussion it is also invariant under generalised (Abelian) gauge transformations (5.5). In terms of $V(z)$ it may be written as (Fayet–Illiopoulos term)

$$I_D = \varkappa \int d^4x D(x) = 2\varkappa \int d^8z V \tag{5.22}$$

where we use (5.6) and (4.21).

The quadratic mass term (ignoring θ-dependent surface terms)

$$\int d^4x \left(\frac{D^\alpha \bar{D}^2 D_\alpha}{8}\right) V^2 = 2 \int d^8z V^2 \tag{5.23}$$

discussed above is supersymmetric but fails to be invariant under (5.4). We remark that if v_m is a polar vector then D is a pseudoscalar so that (5.22) violates parity.

5.6 Supersymmetric Abelian gauge theory

5.6.1 U(1) Gauge transformation of chiral superfield

Consider a gauge group $U(1)_q$ under which a set of matter superfields Φ_i transform as

$$\Phi_i \to e^{-i\Lambda q_i}\Phi_i \qquad \bar{\Phi}_i \to \bar{\Phi}_i e^{i\bar{\Lambda} q_i} \tag{5.24}$$

where Λ and $\bar{\Lambda}$ are defined in (5.2) and q_i are real charges. The chiral superfield is mapped to a chiral superfield and the Φ_i carry a representation of the $U(1)_q$ gauge group. The kinetic term, however, is not invariant under (5.24)

$$\bar{\Phi}_i\Phi_i \to \bar{\Phi}_i e^{i\bar{\Lambda} q_i} e^{-i\Lambda q_i}\Phi_i. \tag{5.25}$$

Just like in ordinary gauge theory we must introduce a compensating gauge superfield to restore this invariance. In this process we obtain also the minimal coupling of the gauge superfield with the matter superfields. For the U(1) gauge group it is suggested from (5.4) and (5.5) that we introduce a corresponding compensating vector superfield $V(z)$ and consider (5.4) and (5.24) as the SS generalisation of gauge transformations for the matter and gauge supermultiplets. To restore the generalised gauge invariance we may

† Fayet and Illiopoulos (1974).

modify the kinetic term to the form

$$\bar{\Phi}_i e^{g_1 q_i V} \Phi_i \tag{5.26}$$

which is invariant under simultaneous transformations (5.4) and (5.24).

To construct an SS invariant from (5.26) we observe that (5.26) satisfies the reality constraint like V. The space–time integral of the D-term of (5.26) is an SS invariant. From the expression of the D-term in terms of covariant derivatives in (5.6) and ignoring the surface terms we may write the modified kinetic term for each Φ_i in the SS action as

$$\int d^8z \bar{\Phi}_i e^{g_1 q_i V} \Phi_i = \int d^4x \bar{\Phi}_i e^{g_1 q_i V} \Phi_i \,|_{\theta^2\bar{\theta}^2}. \tag{5.27}$$

This expression contains the usual kinetic energy terms for the matter superfield along with the minimal interaction terms of a gauge supermultiplet.

5.6.2 Supersymmetric generalised gauge invariant action. Wess–Zumino gauge

The general SS action for U(1) gauge theory may be written† (apart from a gauge-fixing term) as

$$I = \sum_i \int d^8z \bar{\Phi}_i e^{g_1 q_i V} \Phi_i + \left(\int d^6 s W(\Phi) + \int d^6 \bar{s} \bar{W}(\bar{\Phi}) \right) + I_V + I_D \tag{5.28}$$

where I_V and I_D are given in (5.21) and (5.22) respectively and $W(\Phi)$ is the SS superpotential. There is a constraint on the general form of W arising from U(1) invariance.

The exponential in V gives rise to an infinite series of interaction terms involving V. However, if we use the non-supersymmetric *Wess–Zumino gauge* we get a finite number of terms. This choice of gauge sets $C = M = N = \chi = \bar{\chi} = 0$. The possibility of realising it (at least at the classical level) is evident from the generalised gauge transformations in (5.5). The superfield V then takes the simple form

$$V(x, \theta, \bar{\theta}) = -(\theta\sigma^l\bar{\theta})v_l(x) + i\theta^2\bar{\theta}\bar{\lambda}(x) - i\bar{\theta}^2\theta\lambda(x) + \tfrac{1}{2}\theta^2\bar{\theta}^2 D(x). \tag{5.29}$$

It is easily shown that in this gauge

$$V^2 = -\tfrac{1}{2}\theta^2\bar{\theta}^2 v^l v_l \tag{5.30}$$

$$V^3 = 0. \tag{5.31}$$

The gauge still allows the Abelian gauge transformations $v_l \to v_l + \partial_l\omega$.

† A term involving integration over the whole of superspace (d^8z) is called a D-term while those involving d^6s or $d^6\bar{s}$ are called F-terms.

In the Wess–Zumino gauge the component form of (5.28) is found to be (denoting the coefficient of $\theta^2\bar{\theta}^2$ by $|_{\theta^2\bar{\theta}^2}$)

$$\int d^8z\bar{\Phi}e^{g_1qV}\Phi \equiv \int d^4x\bar{\Phi}e^{g_1qV}\Phi\,|_{\theta^2\bar{\theta}^2}$$

$$= \int d^4x\left(\bar{\Phi}\Phi + g_1\bar{\Phi}qV\Phi + \frac{g_1^2}{2}\bar{\Phi}q^2V^2\Phi\right)\Bigg|_{\theta^2\bar{\theta}^2}$$

$$= \int d^4x\Big(-(\bar{\mathscr{D}}_l\bar{A})(\mathscr{D}^lA) - i\psi\sigma^l\bar{\mathscr{D}}_l\bar{\psi} + \bar{F}F \qquad (5.32)$$

$$+\frac{ig_1}{\sqrt{2}}(\bar{A}\psi\lambda - A\bar{\psi}\bar{\lambda})q + \frac{1}{2}g_1\bar{A}qAD\Big)$$

$$I_V + I_D = \int d^4x(-\tfrac{1}{4}F^{lm}F_{lm} - i\lambda\sigma^l\partial_l\bar{\lambda} + \tfrac{1}{2}D^2 + \varkappa D)$$

where we have suppressed the index i for convenience and $\mathscr{D}_l^i = (\partial_l + (ig_1/2)q_iv_l)$, $\bar{\mathscr{D}}_l^i = (\partial_l - (ig_1/2)q_iv_l)$ are ordinary gauge covariant derivatives. Recalling the discussion in Chapter 4 the equations of motion for the auxiliary fields are

$$D = -(\varkappa + \tfrac{1}{2}g_1\bar{A}YA) = -(\varkappa + \tfrac{1}{2}g_1\bar{A}_iq_iA_i)$$
$$\bar{F}_i = -\frac{\partial W}{\partial A_i} \qquad F_i = -\frac{\partial \bar{W}}{\partial \bar{A}_i} \qquad (5.33)$$

where Y indicates the U(1) charge matrix of chiral superfields. On eliminating the auxiliary fields we obtain for the scalar potential

$$V = \sum_i |\bar{F}_iF_i| + \tfrac{1}{2}D^2 \geqslant 0. \qquad (5.34)$$

We shall see in the next chapter that a non-vanishing value of $\varkappa$ in some cases may lead to spontaneous SS breaking.

For the discussion in the next chapter we anticipate here the result for non-Abelian gauge theory (Chapter 7). The superfields Φ_i now carry a representation of some semi-simple non-Abelian gauge group with Hermitian generators $(T^a)_{ij}$ where a is the group index†. The complete scalar potential then may be written as

$$V = \sum_i \bar{F}_iF_i + \sum \tfrac{1}{2}(D^a)^2 + \sum \tfrac{1}{2}D^2 \geqslant 0 \qquad (5.35)$$

where $D^a = \frac{1}{2}g\bar{A}_i(T^a)_{ij}A_j$ and the sum in the second term on the RHS is

† The gauge superfields V^a transform in the adjoint representation and satisfy $V^a = \bar{V}^a$. See Chapter 7.

over the several groups contained in the semi-simple group and the various chiral superfield representations while in the last term the sum is over all the U(1) factors.

5.6.3 *Supersymmetric QED. R-symmetry in gauge theory*

The SS extension of QED is obtained if we have two superfields denoted by Φ_+ and Φ_- carrying $U(1)_{em}$ charges $+1$ and -1. The action is ($g_1 = e$):

$$I = \int d^8z(\bar{\Phi}_+ e^{eV}\Phi_+ + \bar{\Phi}_- e^{-eV}\Phi_-) + m\left(\int d^6s\Phi_+\Phi_- + \text{cc}\right) + I_V. \tag{5.36}$$

The Lagrangian written in the Wess–Zumino gauge reads†

$$\begin{aligned}\mathscr{L} = &-(\partial^l\bar{A}_+)(\partial_lA_+) - (\partial^l\bar{A}_-)(\partial_lA_-) - i\psi_+\sigma^l\partial_l\bar{\psi}_+ \\ &- i\psi_-\sigma^l\partial_l\bar{\psi}_- + \bar{F}_+F_+ + \bar{F}_-F_- + \frac{ie}{2}v_l[\bar{A}_+\partial_lA_+ \\ &- (\partial_l\bar{A}_+)A_+ - \bar{A}_-\partial_lA_- + (\partial_l\bar{A}_-)A_-] - \frac{e}{2}v_l(\psi_+\sigma_l\bar{\psi}_+ - \psi_-\sigma_l\bar{\psi}_-) \\ &- \frac{e^2}{4}v_lv^l(\bar{A}_+A_+ + \bar{A}_-A_-) + \frac{e}{2}D(\bar{A}_+A_+ - \bar{A}_-A_-) \\ &+ \frac{ie}{\sqrt{2}}(\bar{A}_+\psi_+\lambda - A_+\bar{\psi}_+\bar{\lambda} - \bar{A}_-\psi_-\lambda + A_-\bar{\psi}_-\bar{\lambda}) \\ &+ m(A_+F_- + A_-F_+ + \bar{A}_+\bar{F}_- + \bar{A}_-\bar{F}_+ - \psi_+\psi_- \\ &- \bar{\psi}_+\bar{\psi}_-) + \mathscr{L}_V.\end{aligned} \tag{5.37}$$

The 2-spinors λ, $\bar{\lambda}$ give rise to a Majorana 4-spinor so that in 4-component notation

$$\mathscr{L}_V = -\frac{1}{4}F^{lm}F_{lm} + \frac{i}{2}\bar{\lambda}\gamma^l\partial_l\lambda + \frac{1}{2}D^2 \tag{5.38}$$

where the Majorana spinor

$$\lambda = \begin{pmatrix}\lambda_\alpha \\ \bar{\lambda}^{\dot{\alpha}}\end{pmatrix}$$

is identified with the SS partner of the photon called the photino. Considering, say, the minimal interaction term involving v_l we conclude that the massive Dirac spinor for the charged electron must be identified as

$$\psi = \begin{pmatrix}\psi_{+\alpha} \\ \bar{\psi}_-^{\dot{\alpha}}\end{pmatrix} \qquad \bar{\psi} = i(\psi_-^\alpha, \bar{\psi}_{+\dot{\alpha}}). \tag{5.39}$$

† Wess and Zumino (1974c).

We then find

$$\mathscr{L} = \mathrm{i}\bar{\psi}(\gamma_l \mathscr{D}^l_+ + m)\psi - (\bar{\mathscr{D}}^l_+ \bar{A}_+)(\mathscr{D}_{+l} A_+) - (\bar{\mathscr{D}}^l_- \bar{A}_-)(\mathscr{D}_{-l} A_-)$$
$$+ \bar{F}_+ F_+ + \bar{F}_- F_- + m(A_+ F_- + A_- F_+ + \bar{A}_+ \bar{F}_-$$
$$+ \bar{A}_- \bar{F}_+) + \frac{e}{2} D(\bar{A}_+ A_+ - \bar{A}_- A_-) + \frac{e}{2\sqrt{2}} [(\bar{A}_+ + A_-)\bar{\lambda}\psi$$
$$+ (A_- - \bar{A}_+)\bar{\lambda}\mathrm{i}\gamma_5\psi - (A_+ + \bar{A}_-)\bar{\psi}\lambda + (\bar{A}_- - A_+)\bar{\psi}\mathrm{i}\gamma_5\lambda] + \mathscr{L}_V \quad (5.40)$$

where $\mathscr{D}_l^{\pm} = (\partial_l \pm (\mathrm{i}e/2(v_l)$ and $\bar{\mathscr{D}}_l^{\pm} = (\partial_l \mp (\mathrm{i}e/2)v_l)$.

There is a Yukawa coupling of electron, photino and charged scalar fields and a quartic coupling of scalar fields following from (5.33) which gives $D = -(e/2)(\bar{A}_+ A_+ - \bar{A}_- A_-)$, $\bar{F}_+ = -mA_-$ and $\bar{F}_- = -mA_+$. The matter fields A_+, A_- and ψ carry the same mass while the photino is massless.

The $\mathrm{U}(1)_{\mathrm{em}}$ gauge invariance of (5.36) implies electromagnetic charges for $A_{\pm}$, ψ. The photino has no charge and does not interact directly with the photon. We may in addition identify an R symmetry in (5.36) giving rise to a $\mathrm{U}(1)_{\mathrm{R}}$ global symmetry. To see this we assign the following R charges to the superfields:

$$R(\Phi_{\pm}) = -R(\bar{\Phi}_{\pm}) = 1 \qquad R(V) = 0. \quad (5.41)$$

The R charges of the component fields are then derived from $R(\theta) = -R(\bar{\theta}) = 1$. We obtain

$$R(C) = R(D) = R(v_l) = 0 \qquad R(\lambda) = -R(\bar{\lambda}) = 1$$
$$R(\chi) = -R(\bar{\chi}) = -1 \qquad R(M + \mathrm{i}N) = -2 \qquad R(M - \mathrm{i}N) = 2 \quad (5.42)$$

and

$$R(A) = -R(\bar{A}) = 1 \qquad R(\psi) = R(\bar{\psi}) = 0 \qquad R(F) = -R(\bar{F}) = -1. \quad (5.43)$$

The invariance (in the Wess–Zumino gauge) is immediately seen from (5.37). Alternatively, we may assign† $R(\mathrm{d}^2\theta) = -R(\mathrm{d}^2\bar{\theta}) = -2$ and consider (5.36). So far as the spinor fields are concerned the R-transformation considered here is simply the well known chirality transformation of the Majorana field. The conserved $\mathrm{U}(1)_{\mathrm{R}}$ Noether current is easily constructed:

$$j^l_{\mathrm{R}} = \delta\lambda^{\alpha} \frac{\delta\mathscr{L}}{\delta(\partial_l \lambda^{\alpha})} + \delta\bar{\lambda}_{\dot{\alpha}} \frac{\delta\mathscr{L}}{\delta(\partial_l \bar{\lambda}_{\dot{\alpha}})} + \delta A_+ \frac{\delta\mathscr{L}}{\delta(\partial_l A_+)} + \delta\bar{A}_+ \frac{\delta\mathscr{L}}{\delta(\partial_l \bar{A}_+)} + \dots$$
$$= -\lambda\sigma^l\bar{\lambda} + \mathrm{i}(\bar{A}_+ \mathscr{D}^l_+ A_+ - A_+ \bar{\mathscr{D}}^l_+ \bar{A}_+) + \mathrm{i}(\bar{A}_- \mathscr{D}^l_- A_- - A_- \bar{\mathscr{D}}^l_- \bar{A}_-) \quad (5.44)$$

† $\int \mathrm{d}^2\theta\theta^2 = 1$ implies $R(\mathrm{d}^2\theta) = -2$ etc. Similarly, $R(\mathrm{D}_{\alpha}) = R(\partial_{\alpha}) = -1$, $R(\bar{\mathrm{D}}_{\dot{\alpha}}) = 1$, etc.

where the first term is a (axial vector) chiral current since $2\lambda\sigma^l\bar{\lambda} = \bar{\lambda}\mathrm{i}\gamma_5\gamma^l\lambda$ and $\bar{\lambda}\gamma^l\lambda = 0$. The conserved EM current corresponding to global $U(1)_{\mathrm{em}}$ invariance on the other hand is

$$j^l_{\mathrm{em}} = \psi_-\sigma^l\bar{\psi}_- - \psi_+\sigma^l\bar{\psi}_+ + \mathrm{i}(\bar{A}_+\mathscr{D}^l_+A_+ - A_+\bar{\mathscr{D}}^l_+\bar{A}_+)$$
$$- \mathrm{i}(\bar{A}_-\mathscr{D}^l_-A_- - A_-\bar{\mathscr{D}}^l_-\bar{A}_-). \tag{5.45}$$

We remark that the Wess–Zumino gauge is non-supersymmetric. The modified SS transformations which leave the action (5.37) invariant are nonlinear and will be discussed in §7.3 along with the case of non-Abelian gauge theory. For the Abelian case they are given in (7.45) and (7.48). The 'supersymmetry' Noether current may then be constructed (see (2.33)):

$$(\xi J^l + \bar{\xi}\bar{J}^l) = \delta v^m \frac{\delta\mathscr{L}}{\delta(\partial_l v^m)} + \delta\lambda \frac{\delta\mathscr{L}}{\delta(\partial_l\lambda)} + \delta\psi_+ \frac{\delta\mathscr{L}}{\delta(\partial_l\psi_+)} + \ldots - K^l \tag{5.46}$$

where $\delta\mathscr{L} = \partial_l K^l$. We find

$$J^l = \tfrac{1}{2}\sigma^l[\bar{\lambda}D - \mathrm{i}\sqrt{2}\,\bar{\psi}_+F_+ - \mathrm{i}\sqrt{2}\,\bar{\psi}_-F_-] + \ldots$$
$$\bar{J}^l = -\tfrac{1}{2}\bar{\sigma}^l[\lambda D + \mathrm{i}\sqrt{2}\,\psi_+\bar{F}_+ + \mathrm{i}\sqrt{2}\,\psi_-\bar{F}_-] + \ldots \tag{5.47}$$

where the dots indicate bilinear and trilinear terms and they are independent of the auxiliary fields.

Problems

5.1 Show that

$$\mathrm{e}^{-\mathrm{i}\Lambda} = U\mathrm{e}^{-\mathrm{i}\tilde{A}}[1 - \mathrm{i}\sqrt{2}\,\theta\tilde{\psi} + \theta^2(\tfrac{1}{2}\tilde{\psi}\tilde{\psi} - \mathrm{i}\tilde{F})]$$
$$\mathrm{e}^{-\mathrm{i}\Lambda}\Phi = U\mathrm{e}^{-\mathrm{i}\tilde{A}}[A + \sqrt{2}\,\theta(\psi - \mathrm{i}A\tilde{\psi}) + \theta^2(F - \mathrm{i}\tilde{F}A + \mathrm{i}\tilde{\psi}\psi + \tfrac{1}{2}A\tilde{\psi}\tilde{\psi})].$$

5.2 Show that

$$V = V_{\mathrm{WZ}} + \mathrm{i}(\bar{\Lambda}_0 - \Lambda_0)$$

where

$$\Lambda_0 = U\left(\frac{\mathrm{i}C}{2} - \theta\chi - \frac{1}{2}(M + \mathrm{i}N)\theta^2\right).$$

V_{WZ} is given in (5.29).

5.3 Verify that

$$e^{-i\Lambda_0}\Phi = Ue^{C/2}\left[A+\sqrt{2}\,\theta\left(\psi+\frac{i}{2}A\chi\right)\right.$$
$$\left.+\theta^2\left(F+\frac{i}{2}A(M+iN)-\frac{i}{2}\chi\psi+\frac{1}{4}A\chi\chi\right)\right].$$

Compute the SS transformation of the coefficient of θ^2.

5.4 Show that the Wess–Zumino gauge is non-supersymmetric.

5.5 Establish the last relation in (5.6).
Hints:

$$\begin{aligned}D_\alpha\bar{D}^2D_\beta &= (\{D_\alpha,\bar{D}_{\dot\alpha}\}-\bar{D}_{\dot\alpha}D_\alpha)(\{\bar{D}^{\dot\alpha},D_\beta\}-D_\beta\bar{D}^{\dot\alpha})\\ &= -4\partial_{\alpha\dot\alpha}\partial_\beta{}^{\dot\alpha}+2i(\partial_{\alpha\dot\alpha}D_\beta\bar{D}^{\dot\alpha}+\partial_\beta{}^{\dot\alpha}\bar{D}_{\dot\alpha}D_\alpha)+\tfrac{1}{2}\bar{D}_{\dot\alpha}D^2\bar{D}^{\dot\alpha}\end{aligned}$$
$$\bar{D}D^2\bar{D} = D\bar{D}^2D = [\partial^2\bar{\partial}^2-2\partial^\alpha{}_{\dot\alpha}\partial_\alpha{}^{\dot\alpha}+\theta\text{-dependent terms}].$$

5.6 Verify the expressions for δD, δv_l in (5.7).
Hints:

$$\begin{aligned}+8\delta D &= \xi_\beta D^\beta D^\alpha\bar{D}^2D_\alpha V| + \mathrm{CC} = \tfrac{1}{2}\xi^\alpha D^2[\bar{D}^2,D_\alpha]V| + \mathrm{CC}\\ &= 2i\xi^\alpha D^2\bar{D}^{\dot\alpha}\partial_{\alpha\dot\alpha}V| + \mathrm{CC}\end{aligned}$$
$$-2\delta v_{\alpha\dot\alpha} = [\xi_\alpha D^2\bar{D}_{\dot\alpha}-\bar{\xi}_{\dot\alpha}\bar{D}^2D_\alpha+2i(\xi D+\bar{\xi}\bar{D})\partial_{\alpha\dot\alpha}]V|.$$

5.7 Compute $D_\alpha V$ and $U^{-1}D_\alpha V$.

5.8 Show that $W^2 = W^\alpha W_\alpha$

$$W^2| = -16\lambda\lambda \qquad D^\alpha W^2| = -16i[2\lambda^\alpha D - i(\lambda\sigma^l\bar{\sigma}^m)^\alpha F_{lm}]$$

while $D^2W^2|$ is given by (5.18).

5.9 Show that (at the classical theory level)

$$\bar{D}^2(\bar{\Phi}e^V\Phi)| = -4A\left(\bar{F}+\frac{i}{\sqrt{2}}\bar{\psi}\bar{\chi}+\frac{1}{4}\bar{A}\bar{\chi}\bar{\chi}-\frac{i}{2}(M-iN)\bar{A}\right)e^C.$$

The RHS is gauge invariant under (5.5) and does not contain λ, D or F_{lm}. In quantised theory the local product of operators like $\bar{\Phi}\,e^V\Phi$ must be defined carefully and we then obtain anomalous terms on the RHS involving the components of W^α.

5.10 Show that supersymmetry forbids an anomalous magnetic moment $(g-2)$ for a lepton. (See Ferrara and Remiddi (1974).) This is still a current research problem in the context of supergravity (Chapter 11).

6

SPONTANEOUS SUPERSYMMETRY BREAKING

6.1 Introduction

A symmetry is spontaneously broken if the vacuum or ground state of the theory is not left invariant under the operation of the symmetry. For realistic applications in high energy physics supersymmetry must be spontaneously broken. For example, this is desirable so as to make an unobserved superpartner highly massive, analogous to the case of spontaneously broken ordinary gauge theories.

Supersymmetry is spontaneously broken when some of the auxiliary fields in the SS theory acquire non-vanishing vacuum expectation values (VEV). This may be seen, for example, from (4.8). If† $\langle F\rangle_0 = f$, $\langle A\rangle_0 = a$ where a, f are constant due to translation invariance, then we find‡

$$\langle\delta\psi_\alpha\rangle_0 = \mathrm{i}\langle\,[\xi Q + \bar{\xi}\bar{Q}, \psi_\alpha]\,\rangle_0 = -\sqrt{2}\,\xi_\alpha f. \tag{6.1}$$

The generators $Q_\alpha, \bar{Q}_{\dot\alpha}$ thus do not all annihilate the vacuum state $|0\rangle$ and supersymmetry is spontaneously broken. We note also that a non-vanishing value of $\langle A\rangle_0$ alone does not lead to broken rigid supersymmetry. An analogous discussion based on (5.7) shows that a non-vanishing vacuum expectation value (VEV) for the auxiliary field D would also imply a broken supersymmetry.

An alternative discussion of the realisation of global or rigid supersymmetry in the spontaneously broken mode may be based on the result (1.78) following as a consequence of the $N=1$ supersymmetry algebra. It follows that in a theory with global supersymmetry $\langle H\rangle \geqslant 0$ for every state. The vacuum or the lowest energy state (or the family of states), $|0\rangle$, will have vanishing energy if and only if all Q_α and $\bar{Q}^{\dot\alpha}$ annihilate it. The states with vanishing energy density are the supersymmetric ground states of the theory. If the supersymmetry is spontaneously broken, at least one of the SS generators does not annihilate the vacuum state $|0\rangle$ and the ground state energy is not exactly zero. The vacuum energy is thus the order parameter for spontaneously broken global supersymmetry (see figure 6.1). It follows from the expression (5.35) for the scalar potential that the broken

† Spinor and tensor fields must have vanishing VEV due to rotational invariance.
‡ It is clear from the context that the ψ_α, Q_α etc written here are field operators.

supersymmetry occurs if (*a*) $F_i = 0$ have no solution: *F*-type breaking, (*b*) $D = 0, D^a = 0$ have no solution: *D*-type breaking, (*c*) $F_i = 0, D = 0, D^a = 0$ have solutions but they do not have a common solution. In a theory with no spontaneously broken supersymmetry (global) the set of equations $F_i = 0, D = 0, D^a = 0$ have a common solution and give rise to an SS ground state (or states) with zero energy. Analogous to the appearance of the Goldstone boson there appears in the spontaneously broken supersymmetric theory a massless spin-$\frac{1}{2}$ Goldstone fermion or goldstino which carries an inhomogeneous term in its supersymmetry transformation law (see illustrations below). Moreover, the different members of some supermultiplets acquire different masses.

We will say only a few words here about the local supersymmetry (supergravity) and its breaking; it will be described in Chapter 11. When the global supersymmetry is promoted to local supersymmetry the supergravity multiplet $(2, \frac{3}{2})$ mentioned in Chapter 2 enters the theory as the gauge fields of local supersymmetry. The additional gravitational interactions now allow the vacuum energy to be positive, negative or vanishing. In fact in the absence of SS breaking the vacuum energy is found to be negative while a positive value for it indicates broken supergravity. This is in contrast to global supersymmetry where the vacuum energy gets only positive contributions proportional to the SS breaking scale. The local supersymmetry resolves the problem of the apparent non-existence of goldstino particles of broken rigid supersymmetry through the super-Higgs mechanism, analogous to the standard Higgs mechanism of ordinary gauge theories.

6.2 Illustrations

We now give several examples of broken and unbroken supersymmetry.

6.2.1 *Wess–Zumino model*

The superpotential $W(A) = \frac{1}{2}mA^2 + \frac{1}{3}gA^3$ of the model gives

$$-\bar{F} = (mA + gA^2) \qquad M = (m + 2gA) \tag{6.2}$$

$$V = |mA + gA^2|^2. \tag{6.3}$$

The solutions of $\bar{F} = 0$ give the ground states with the following VEV for the physical field A

$$\text{(i) } \langle A\rangle_0 = 0 \qquad \text{(ii) } \langle A\rangle_0 = -\frac{m}{g}. \tag{6.4}$$

The plot of the scalar potential V against Re A show the two minima at these values while a maximum is obtained for $\langle A \rangle_0 = -m/2g$ (see figure 6.1(*c*)).

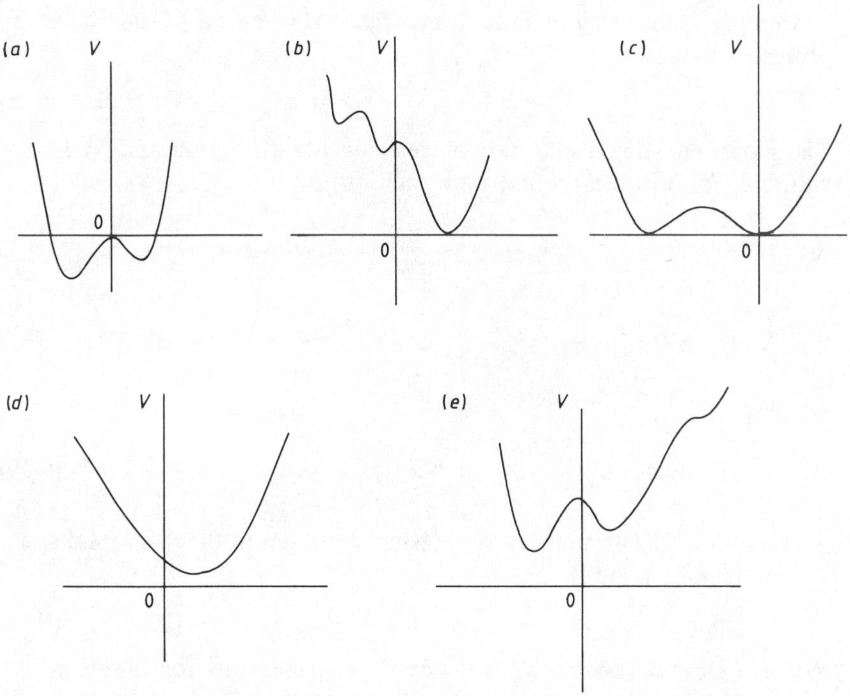

Figure 6.1 (*a*) Non-supersymmetric potential; (*b*) and (*c*) potentials for an unbroken supersymmetry; (*d*) and (*e*) potentials for spontaneously broken supersymmetry.

6.2.2 $W(\Phi) = f\Phi + b, f \neq 0$

This offers itself as the simplest example of spontaneous SS breaking. In fact $\bar{F} = -f$ and $V = |f|^2 > 0$. On making the shift $\bar{F} \to \bar{F} - f$ the SS transformation of the spinor component acquires an inhomogeneous term identifying it as the Goldstone fermion.

6.2.3 *Model with an internal symmetry*

Three chiral superfields Φ_+, Φ_-, Φ with

$$W(A) = gAA_+A_- - \mu A + \tfrac{1}{6} fA^3. \tag{6.5}$$

The action here possesses an internal U(1) symmetry

$$\Phi_+ \to e^{i\lambda\psi}\Phi_+ \qquad \Phi_- \to e^{-i\lambda\psi}\Phi_- \qquad \Phi \to \Phi \tag{6.6}$$

while W by itself is invariant under the complex extension of U(1)

$$A_+ \to e^{\lambda\psi} A_+ \qquad A_- \to e^{-\lambda\psi} A_- \qquad A \to A. \tag{6.7}$$

This larger symmetry of the superpotential has interesting consequences concerning SS breaking by radiative corrections when gauge superfields are present. From (4.8) we derive

$$-\bar{F}_+ = gA_-A \qquad -\bar{F}_- = gA_+A \qquad -\bar{F} = gA_+A_- - \mu + \tfrac{1}{2} fA^2. \tag{6.8}$$

The supersymmetry is unbroken since a consistent solution of $F_i = 0$ exists with the following SS ground state solutions

$$\text{(i)}\ A_+ = A_- = 0 \qquad A^2 = \frac{2\mu}{f} \qquad \text{U(1) symmetric}$$

$$\text{(ii)}\ A = 0,\ A_+A_- = \frac{\mu}{g} \qquad \text{U(1) spontaneously broken} \tag{6.9}$$

giving an infinitely degenerate SS vacuum

$$A = 0 \qquad A_+ = \sqrt{\frac{\mu}{g}}\, e^{\alpha} \qquad A_- = \sqrt{\frac{\mu}{g}}\, e^{-\alpha}. \tag{6.10}$$

For $f = 0$ the U(1) symmetry is necessarily broken in the SS ground state. The potential energy is

$$V = g^2 |A_+A|^2 + g^2 |A_-A|^2 + |gA_+A_- - \mu + \tfrac{1}{2} fA^2|^2 \tag{6.11}$$

We also find† det $\mathbf{M} \not\equiv 0$ while det $\mathbf{M}|_{\min}$ vanishes only for case (ii).

6.2.4 *O'Raifeartaigh model*‡

We discuss in detail a model with three chiral superfields with scalar components X, Y and Z and interaction given by the superpotential

$$W = \lambda X(Z^2 - M^2) + gYZ. \tag{6.12}$$

We have

$$-\bar{F}_x = \lambda(Z^2 - M^2) \qquad -\bar{F}_y = gZ \qquad -F_z = 2\lambda XZ + gY. \tag{6.13}$$

Since $F_x = 0$ leads to $Z = M$ and $F_y \neq 0$ the SS is clearly broken. The potential is given by

$$V(X, \bar{X}, Y, \bar{Y}, Z, \bar{Z}) = \lambda^2 |Z^2 - M^2|^2 + g^2 |Z|^2 + |2\lambda XZ + gY|^2 > 0. \tag{6.14}$$

† $\mathbf{M}$ indicates the fermion mass matrix.

‡ O'Raifeartaigh (1975).

Its minimum is obtained from

$$V_x = \frac{\partial V}{\partial X} = 2\lambda Z(2\lambda \bar{X}\bar{Z} + g\bar{Y}) = 0$$

$$V_y = \frac{\partial V}{\partial Y} = g(2\lambda \bar{X}\bar{Z} + g\bar{Y}) = 0 \tag{6.15}$$

$$V_z = \frac{\partial V}{\partial Z} = 2\lambda^2 Z(\bar{Z}^2 - M^2) + g^2\bar{Z} + 2\lambda X(2\lambda \bar{X}\bar{Z} + g\bar{Y}) = 0$$

or

$$\begin{aligned} 2\lambda \bar{X}\bar{Z} + g\bar{Y} &= 0 \\ 2\lambda^2 Z(\bar{Z}^2 - M^2) + g^2\bar{Z} &= 0. \end{aligned} \tag{6.16}$$

Solving these equations we find in the ground state, $(2\lambda^2 M^2 < g^2)$,

$$\begin{array}{lll} \langle Z\rangle_0 = 0 & \langle Y\rangle_0 = 0 & \langle X\rangle_0 \text{ undetermined} \\ \langle F_x\rangle_0 = \lambda M^2 & \langle F_y\rangle_0 = \langle F_z\rangle_0 = 0 & E_{\text{vac}} = V_0 = \lambda^2 M^4 > 0. \end{array} \tag{6.17}$$

The fermion mass matrix $\mathbf{M}$ of §4.2. is given by

$$\mathbf{M} = \begin{pmatrix} 0 & 0 & 2\lambda Z \\ 0 & 0 & g \\ 2\lambda Z & g & 2\lambda X \end{pmatrix} \qquad \det \mathbf{M} \equiv 0. \tag{6.18}$$

The quadratic terms in $\mathscr{L}_{\text{Yuk}}$ (equation(4.30))

$$-g(\psi_y\psi_z + \bar{\psi}_y\bar{\psi}_z) = \mathrm{i}g\bar{\psi}\psi \tag{6.19}$$

where

$$\psi = \begin{pmatrix} \psi_{y\alpha} \\ \bar{\psi}_z^{\dot{\alpha}} \end{pmatrix}$$

indicate a Dirac particle of mass g while the fermion ψ_x is massless; it is the Goldstone fermion. The shift $X \to X + \langle X\rangle_0$ does not affect the mass of ψ_x and we will set $\langle X\rangle_0 = 0$ for convenience.

The relevant non-vanishing elements of the boson mass matrix are

$$\begin{array}{lll} V_{x\bar{x}} = 4\lambda^2 |Z|^2 & V_{y\bar{y}} = g^2 & V_{z\bar{z}} = 4\lambda^2(|Z|^2 + |X|^2) + g^2 \\ V_{zz} = 2\lambda^2(\bar{Z}^2 - M^2) & V_{\bar{z}\bar{z}} = \bar{V}_{zz} \text{ etc.} & \end{array} \tag{6.20}$$

Their values calculated at the minimum $\langle X\rangle_0 = \langle Y\rangle_0 = \langle Z\rangle_0 = 0$ indicate that the scalar field X is massless and Y has mass g. The model thus contains a non-Goldstone spin-zero boson X which remains massless at the classical level and acquires a mass by radiative corrections. The complex scalar field Z shows mass splitting. Writing $Z = Z_1 + \mathrm{i}Z_2$ where Z_1 and Z_2

are real we find

$$\tfrac{1}{2}(V_{0zz}Z^2 + V_{0\bar{z}\bar{z}}\bar{Z}^2 + 2V_{0z\bar{z}}\,|\,Z\,|^2)$$
$$= [(g^2 - 2\lambda^2 M^2)Z_1^2 + (g^2 + 2\lambda^2 M^2)Z_2^2]. \qquad (6.21)$$

Consequently the two real scalars have (mass)^2 given by $(g^2 - 2\lambda M_s^2)$ and $(g^2 + 2\lambda M_s^2)$ where $M_s = \surd\langle F_x\rangle_0 = \surd(\lambda M^2)$ is the SS breaking scale. This is the only place in the mass spectrum where we see that the SS is broken. The splitting arises due to the Z-supermultiplet coupling to the goldstino supermultiplet X in the superpotential with strength λ. The fermions on the other hand do not feel the SS breakdown at the tree level. The mass relation Str $\mathbf{M}^2 = \Sigma(-1)^J(2J+1)\, m_J^2 = 0$ where J indicates the spin of the particle and the summation runs over all the self-charge conjugate particles may be easily verified. For the case $2\lambda^2 M^2 > g^2$ see problem 6.3.

6.2.5 *Fayet model*†

Consider the U(1) guage invariant interaction of an Abelian gauge superfield V with a massless chiral superfield Φ. The F term vanishes (because of U(1) invariance) and from (5.33) and (5.34) on setting $g_1 q = e$

$$D = -(\varkappa + \tfrac{1}{2}e\bar{A}A) \qquad V = \tfrac{1}{2}D^2. \qquad (6.22)$$

Two cases may be distinguished (see figure 6.2).

Case 1: $e\varkappa < 0$; SS unbroken but gauge symmetry broken. We find $D = 0$, $\bar{A}A = -2\varkappa/e = v^2$ and we have an infinitely degenerate SS vacuum. Consider the ground state given by $A = \mathrm{i}v, \bar{A} = -\mathrm{i}v, D = 0$. Performing the shifts $A \to A + \mathrm{i}v, \bar{A} \to \bar{A} - \mathrm{i}v, D \to D$ we can find the mass spectrum by collecting the quadratic terms in the shifted Lagrangian. For boson fields we find them from (5.32) and (5.33) to be

$$-\frac{1}{2}\left(\frac{e^2v^2}{2}\right)\left(\frac{\mathrm{i}(\bar{A} - A)}{\sqrt{2}}\right)^2 - \frac{1}{2}\left(\frac{e^2v^2}{2}\right) v^l v_l \qquad (6.23)$$

where we have fixed the gauge by setting $\bar{A} + A = 0$.

For the fermion fields we obtain

$$\frac{ev}{\sqrt{2}}(\psi\lambda + \bar{\psi}\bar{\lambda}). \qquad (6.24)$$

Thus the spin-1 gauge field, the spin-0 Higgs boson and spin-1/2 Dirac fermion field all acquire the same mass $\sqrt{-e\varkappa}$. This is the supersymmetric generalisation of the Higgs mechanism. A chiral supermultiplet combines with a massless vector multiplet to give a massive supermultiplet.

† Fayet (1976).

Alternatively, we may perform a generalised gauge transformation in (5.28) such that the corresponding gauge superfunction Λ in (5.2) is defined by

$$\begin{aligned} e^{-i\Lambda}\Phi &= i\nu \\ e^{i\bar{\Lambda}}\bar{\Phi} &= -i\nu. \end{aligned} \tag{6.25}$$

Then

$$\bar{\Phi}\, e^{eV}\Phi = \nu^2 e^{e\tilde{V}} \tag{6.26}$$

where $e\tilde{V} = eV + i(\Lambda - \bar{\Lambda})$ and in view of (5.5) it has an expansion similar to that of V with the same gauge invariant components F_{lm}, λ and D. Thus $I_{\tilde{V}} = I_V$ and I_D in (5.28) are unchanged while the gauge interaction term becomes

$$\nu^2 \int d^8Z\, e^{e\tilde{V}} = \nu^2 \int d^8Z\Big(e\tilde{V} + \frac{e^2}{2}\tilde{V}^2 + \ldots\Big). \tag{6.27}$$

In this gauge the chiral superfield is totally gauged away and the action is now expressed in terms of a single massive gauge superfield with mass $m = e\nu/\sqrt{2}$ as is evident from the expansion of V^2 given in (5.16). The spin-0 Higgs field in the new gauge is described by the field $\tilde{C}$ component of the massive gauge superfield $\tilde{V}$.

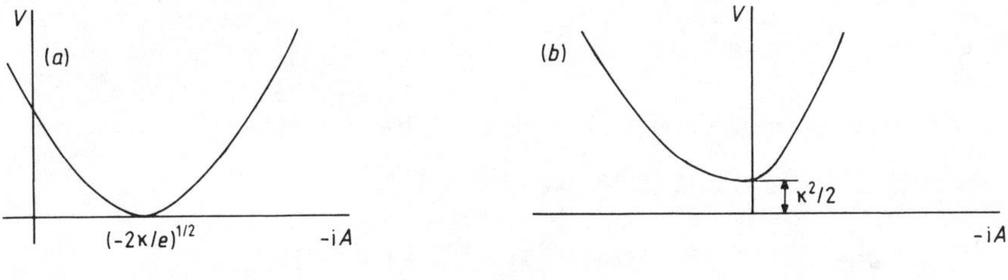

Figure 6.2 Potential in the Fayet model (*a*) Case 1: $e\kappa < 0$, supersymmetry unbroken, U(1) broken. (*b*) Case 2: $e\kappa > 0$ supersymmetry broken, U(1) unbroken.

Eliminating the auxiliary fields $\tilde{M}, \tilde{N}$ and D we may derive to second order in the expansion

$$\begin{aligned} \mathscr{L} = &-\tfrac{1}{4} F_{lm}F^{lm} - i\lambda\sigma^l\sigma_l\bar{\lambda} - (m^2/2)v^l v_l - \tfrac{1}{2}\partial^l(m\tilde{C})\partial_l(m\tilde{C}) - \tfrac{1}{2}m^2(m\tilde{C})^2 \\ &- i(m\chi)\sigma^l\partial_l(m\bar{\chi}) - m[(m\chi)\lambda + (m\bar{\chi})\bar{\lambda}] + \ldots . \end{aligned} \tag{6.28}$$

It shows that we have a spin-1 gauge field v^l, two spin-1/2 fields λ_α and $m\chi_\alpha$

and a spin-0 Higgs field ($m\widetilde{C}$) all having the same mass m. The fermion mass term may be rewritten in two equivalent ways. We may combine λ and $m\chi$ to define a Dirac particle

$$\begin{pmatrix} m\chi^{\alpha} \\ \bar{\lambda}^{\dot{\alpha}} \end{pmatrix}$$

called a 'Dirac-ino', which carries R charge $= -1$. Alternatively, we may define two 'Majorana-inos'

$$\lambda_1 = \frac{(\lambda + m\chi)}{\sqrt{2}} \qquad \lambda_2 = \frac{(\lambda - m\chi)}{\sqrt{2}\,\mathrm{i}} \tag{6.29}$$

and obtain

$$-\mathrm{i}(\lambda_1\sigma^l\partial_l\bar{\lambda}_1 + \lambda_2\sigma^l\partial_l\bar{\lambda}_2) - \frac{m}{2}(\lambda_1\lambda_1 + \bar{\lambda}_1\bar{\lambda}_1) - \frac{m}{2}(\lambda_2\lambda_2 + \bar{\lambda}_2\bar{\lambda}_2)$$

in the Lagrangian.

Case 2: $e\varkappa > 0$; SS broken but gauge symmetry unbroken. At the minimum of the potential when e and $\varkappa$ have the same sign we have $A = \bar{A} = 0$ and $V = \frac{1}{2}\varkappa^2$. The SS is broken because D acquires a nonvanishing VEV, $D = -\varkappa$. The corresponding superpartner λ is the Goldstone fermion. The gauge symmetry is not broken and v^l as well as λ remain massless. The fermion ψ remains massless but the complex scalar A receives (mass)^2 $e\varkappa/2$. The two real scalars thus receive a mass shift in the same direction.

6.2.6 *Supersymmetric* QED *with Fayet–Illiopoulos term*†

From (5.33), (5.34) and (5.36) we obtain

$$\bar{F}_{\pm} = -mA_{\mp} \qquad D = -\left(\varkappa + \frac{e}{2}(|A_+|^2 - |A_-|^2)\right)$$

$$V = \tfrac{1}{2}\varkappa^2 + (m^2 + \tfrac{1}{2}e\varkappa)|A_+|^2 + (m^2 - \tfrac{1}{2}e\varkappa)|A_-|^2 + \frac{e^2}{8}(|A_+|^2 - |A_-|^2)^2 \geqslant 0. \tag{6.30}$$

The minima of V are determined by the equations

$$A_{\pm}[(m^2 \pm \tfrac{1}{2}e\varkappa) + (e^2/4)(|A_{\pm}|^2 - |A_{\mp}|^2)] = 0. \tag{6.31}$$

We may distinguish two cases (see figure 6.3).

† Fayet and Illiopoulos (1974).

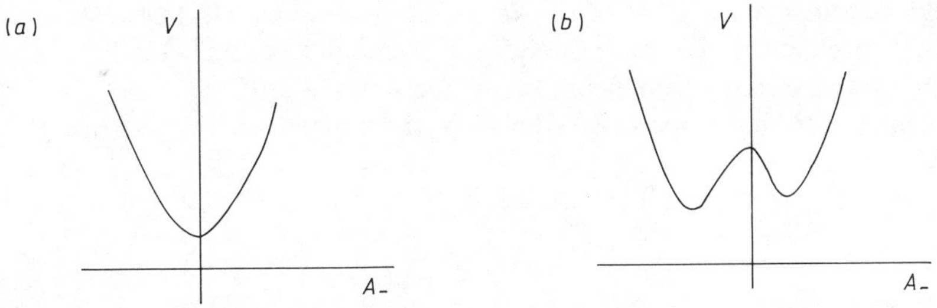

Figure 6.3 Potential in supersymmetric QED. (*a*) Case 1: $m^2 > \frac{1}{2}e\varkappa$. Supersymmetry alone is broken. (*b*) Case 2: $m^2 < \frac{1}{2}e\varkappa$. Both supersymmetry and gauge symmetry are broken.

Case 1: $m^2 > \frac{1}{2}e\varkappa$; $U(1)_{em}$ symmetry unbroken. From (6.31) it follows that $A_+ = A_- = 0$ defines the U(1) symmetric ground state. The supersymmetry is broken because $D = -\varkappa \neq 0$. The corresponding fermion partner λ is the Goldstone fermion. The fermion mass spectrum at the tree level does not feel the broken supersymmetry. However, there is a mass splitting in the boson sector

$$\begin{aligned} m_+^2 &= m^2 + \tfrac{1}{2}e\varkappa \\ m_-^2 &= m^2 - \tfrac{1}{2}e\varkappa \end{aligned} \tag{6.32}$$

revealing a broken supersymmetry.

Case 2: $m^2 < \frac{1}{2}e\varkappa$; $U(1)_{em}$ symmetry broken. To determine a minimum in this case we choose $A_+ = 0$, $A_- = \nu$ where ν may be made real by performing a gauge transformation. From (6.31) we obtain ($\varkappa \neq 0$)

$$(m^2 - \tfrac{1}{2}e\varkappa) + \frac{e^2}{4}\nu^2 = 0 \tag{6.33}$$

and from (6.30), $F_+ = -m\nu$, $F_- = 0$ and $D = -\varkappa + (e/2)\nu^2 = -2m^2/e$. The supersymmetry as well as U(1) gauge symmetry are broken. This is also revealed in the mass spectrum.

We perform the shifts $A_- \to A_- + \nu, A_+ \to A_+, D \to D - \varkappa + (e/2)\nu^2$ and $F_+ \to F_+ - m\nu$ in (5.37) and collect the quadratic mass terms in boson fields

$$-2m^2|A_+|^2 - \frac{1}{2}\left(\frac{e^2\nu^2}{2}\right)\left(\frac{1}{\sqrt{2}}(A_- + \bar{A}_-)\right)^2 - \frac{1}{2}\left(\frac{e^2\nu^2}{2}\right)v^l v_l. \tag{6.34}$$

The shifted Lagrangian is invariant under a modified† U(1) gauge invariance and we may fix the gauge by setting $(A_- - \bar{A}_-) = 0$. The vector

† $i(A_- - \bar{A}_-)$ undergoes an inhomogenous transformation. See, for example, Taylor (1976).

field acquires mass $\surd(\frac{1}{2}e^2v^2) = \surd(e\varkappa - 2m^2)$—broken U(1)—the ordinary Higgs mechanism. We have in addition a complex scalar field A_+ of mass $\surd(2m^2)$ and a real scalar field A_- of mass $\surd(e\varkappa - 2m^2)$.

The quadratic fermion mass terms in the shifted Lagrangian are

$$-m(\psi_+\psi_- + \bar{\psi}_+\bar{\psi}_-) - \frac{iev}{\sqrt{2}}(\psi_-\lambda - \bar{\psi}_-\bar{\lambda})$$
$$= -\psi_-\left(m\psi_+ + \frac{iev}{\sqrt{2}}\lambda\right) + \text{CC}. \tag{6.35}$$

There is a mixing of the fermionic superpartners ψ_+, λ whose corresponding auxiliary fields F_+, D carry non-vanishing VEV's. We must diagonalise (6.35) to read the mass spectrum.

In general if λ^a, λ and ψ_i indicate respectively the fermion superpartners of a non-Abelian gauge supermultiplet, Abelian gauge supermultiplet and chiral supermultiplet we can construct in the theory a massless Majorana spinor called a goldstino $\tilde{\lambda}$ such that it has an inhomogeneous transformation under SS. The desired expression is (see also (5.47))

$$\tilde{\lambda} = -\frac{1}{d}\left[\sum i\langle\bar{F}_i\rangle_0(\sqrt{2}\,\psi_i) + \sum\langle D_a\rangle_0\lambda_a + \sum\langle D\rangle_0\lambda\right]$$
$$\bar{\tilde{\lambda}} = -\frac{1}{d}\left[-\sum i\langle F_i\rangle(\sqrt{2}\,\bar{\psi}_i) + \sum\langle D_a\rangle_0\bar{\lambda}_a + \sum\langle D\rangle_0\bar{\lambda}\right] \tag{6.36}$$

where the dimension of d is (mass)^2 and

$$E_{\text{vac}} = \tfrac{1}{2}d^2 = \sum\langle\bar{F}_i\rangle_0\langle F_i\rangle_0 + \sum\tfrac{1}{2}\langle D_a\rangle_0^2 + \Sigma\tfrac{1}{2}\langle D\rangle_0^2 \tag{6.37}$$

is the vacuum energy density of the spontaneously broken globally SS theory. Calculating $\delta_{SS}\tilde{\lambda}$ we find from (7.45) and (7.55)

$$\langle\delta\tilde{\lambda}_\alpha\rangle_0 = \left(\frac{-1}{d}\right)\left[\sum i\langle\bar{F}_i\rangle_0(-2\langle F_i\rangle_0) + \sum\langle D_a\rangle_0(-i\langle D_a\rangle_0) + \ldots\right]\xi_\alpha$$
$$= +id\xi_\alpha \tag{6.38}$$

Returning to our discussion we have

$$E_{\text{vac}} = \tfrac{1}{2}d^2 = m^2v^2 + \frac{2m^4}{e^2} = \frac{2m^2}{e^2}(e\varkappa - m^2) \tag{6.39}$$

$$\tilde{\lambda} = \frac{-e}{2m\surd(m^2 + \frac{1}{2}e^2v^2)}\left(-i\sqrt{2}\,m\,v\psi_+ - \frac{2m^2}{e}\lambda\right)$$
$$= \frac{1}{\surd(m^2 + \frac{1}{2}e^2v^2)}\left(m\lambda + \frac{iev}{\sqrt{2}}\psi_+\right) \tag{6.40}$$

$$\delta\tilde{\lambda}_\alpha = \frac{2im}{e}\surd(e\varkappa - m^2)\xi_\alpha + \ldots = id\xi_\alpha + \ldots. \tag{6.41}$$

Another linear combination is clearly

$$\tilde{\psi} = \frac{1}{\sqrt{(m^2 + \frac{1}{2}e^2 v^2)}} \left(m\psi_+ + \frac{iev}{\sqrt{2}} \lambda \right) \tag{6.42}$$

where the normalisation is appropriately defined so as to ensure, for example,

$$\psi_+ \sigma^l \partial_l \bar{\psi}_+ + \lambda \sigma^l \partial_l \bar{\lambda} = \tilde{\psi} \sigma^l \partial_l \bar{\tilde{\psi}} + \tilde{\lambda} \sigma^l \partial_l \bar{\tilde{\lambda}}. \tag{6.43}$$

The mass term now becomes

$$-\frac{e}{2m} d(\psi_- \tilde{\psi} + \bar{\psi}_- \bar{\tilde{\psi}}). \tag{6.44}$$

The goldstino is massless while $\tilde{\psi}$ and ψ_- carry mass $(e/2m)d = \sqrt{(e\varkappa - m^2)}$ where d is the supersymmetry breaking scale parameter. Under SS $\tilde{\psi}$ transforms as an ordinary fermion

$$\langle \delta \tilde{\psi}_\alpha \rangle_0 \propto m \langle \delta \psi_+ \rangle_0 + \frac{iev}{\sqrt{2}} \langle \delta \lambda \rangle_0 = -\sqrt{2}\, m \, \langle F_+ \rangle_0 + \frac{ev}{\sqrt{2}} \langle D \rangle_0$$

$$= -2\, m^2 + \frac{ev}{\sqrt{2}} \left(-\frac{2m^2}{e} \right) = 0. \tag{6.45}$$

The conserved supercurrent S^l calculated from the shifted Lagrangian also contains $\bar{\tilde{\lambda}}$:

$$S^l = -\frac{d}{2} \sigma^l \bar{\tilde{\lambda}} + J^l$$

$$\bar{S}^l = +\frac{d}{2} \bar{\sigma}^l \tilde{\lambda} + \bar{J}^l \tag{6.46}$$

where $J^l, \bar{J}^l$ are given in (5.47). In 4-component notation it reads as

$$S^l = \frac{d}{2} \gamma_5 \gamma^l \tilde{\lambda} + J^l.$$

6.2.7 *Non-Abelian gauge theory*

In non-Abelian gauge theories the supersymmetric minima must satisfy

$$\bar{F}_i(A_j) = 0 \tag{6.47}$$

and

$$D^a = \tfrac{1}{2} g \bar{A}_i T^a{}_{ij} A_j = 0. \tag{6.48}$$

The Fayet–Illiopoulos term cannot appear in this case since such a term is not gauge invariant (Chapter 7). It can be shown that if $F_i = 0$ has a solution then we may always find a new solution which satisfies both (6.47) and

(6.48). This derives from the fact that the superpotential $W(\Phi)$ is not only invariant under the gauge group but also under its complex extension. Supersymmetry is spontaneously broken in the non-Abelian case if and only if the equation $\bar{F}_i(A)=0$ has no solution (Wess and Bagger 1983).

Problems

6.1 Discuss the mass spectrum of the supersymmetric model (6.5) for $f=0$ where the U(1) internal symmetry is spontaneously broken. Consider the ground state $\langle A_+\rangle_0=\langle A_-\rangle_0=(\mu/g)^{1/2}$, $\langle A\rangle_0=0$ and verify that because of the supersymmetry the Nambu–Goldstone boson of the broken U(1) appears accompanied by a massless pseudo- or non-Goldstone scalar and a massless non-Goldstone chiral fermion which form a supermultiplet. Hints: Consider the combinations $(\Phi_+\pm\Phi_-)$. Note that a Goldstone particle has an inhomogeneous transformation law.

6.2 Write for the model in problem 6.1 the supercurrent and U(1) current along with their conservation laws.

6.3 For the model (6.12) with spontaneous SS breaking show that

(a) for $2\lambda^2M^2>g^2$ where the parameters are real there exists an asymmetric minimum with

$$\langle Z\rangle_0=\left(M^2-\frac{g^2}{2\lambda^2}\right)^{\frac{1}{2}} \qquad \langle Y\rangle_0=-\frac{2\lambda}{g}\langle Z\rangle_0\langle X\rangle_0 \qquad \langle F_z\rangle_0=0$$

$$\langle F_y\rangle_0=-g\langle Z\rangle_0 \qquad \langle F_x\rangle_0=\frac{g^2}{2\lambda} \qquad E_{\text{vac}}=\tfrac{1}{2}d^2=g^2\left(M^2-\frac{g^2}{4\lambda^2}\right);$$

(b) the model contains a spin$-1/2$ goldstino $\tilde{\lambda}(x)$

$$\tilde{\lambda}(x)=\frac{-\mathrm{i}}{d}\sum\langle\bar{F}_i\rangle_0\sqrt{2}\,\psi_i \qquad \delta_\xi\tilde{\lambda}_\alpha(x)=\mathrm{i}d\xi_\alpha+\ldots$$

which is massless and is accompanied by a massless pseudo-or non-Goldstone scalar $\tilde{A}(x)\approx\sum\langle\bar{F}_i\rangle_0A_i$. Verify that $\tilde{\lambda}(x)$ and $\tilde{A}(x)$ remain massless for any value of the undetermined value $\langle X\rangle_0$.

Construct the Noether supercurrent and write the equation of motion for the goldstino.

Hints: Define $\tilde{\lambda}(x)\sim\langle\bar{F}_x\rangle_0\psi_x+\langle\bar{F}_y\rangle_0\psi_y \qquad \tilde{\psi}(x)\sim\langle F_y\rangle_0\psi_x-\langle F_x\rangle_0\psi_y$

such that

$$\tilde{\lambda}\sigma^l\partial_l\bar{\tilde{\lambda}}+\tilde{\psi}\sigma^l\partial_l\bar{\tilde{\psi}}=\psi_x\sigma^l\partial_l\bar{\psi}_x+\psi_y\sigma^l\partial_l\bar{\psi}_y$$

and $\langle\delta_\xi\tilde{\psi}\rangle_0=0$.

The pseudo-Goldstone gets mass by radiative corrections.

6.4 Work out the details of the models in §6.2.5 and §6.2.6.

6.5 Verify for the models discussed in this chapter the mass relation at the tree level

$$\mathrm{Str}\, \mathbf{M}^2 = \sum_J (-1)^J (2J+1) m_J^2 = 2\, \mathrm{Tr}\, \mathbf{Y} \langle D \rangle_0$$

where J denotes the spin of the particles and $\mathbf{Y}$ indicates the U(1) charge matrix of chiral superfields while $\langle D \rangle_0$ is the VEV of the auxiliary gauge field.

7

NON-ABELIAN SUPERSYMMETRIC GAUGE THEORY

7.1 Gauging of internal symmetry

In §5.6 we discussed the gauging of U(1) internal symmetry in SS theories expressed in the language of superfields. We turn our attention now to the case of an internal symmetry associated with a non-Abelian and compact Lie group $G \otimes \mathrm{U}(1)^p$ where G is semi-simple†. The set of chiral superfields $(\Phi_1, \ldots \Phi_i, \ldots \Phi_n)$ now carries a representation R, in general reducible, of G and the U(1) charges q_k, $k = 1, \ldots, p$. We may choose a set of Hermitian generators T_a of the Lie group G which span a Lie algebra under the commutator (see (1.91))

$$[T_a, T_b] = \mathrm{i} f_{abc} T_c \tag{7.1}$$

where 'a' labels the group index and f_{abc} are real and completely antisymmetric. In any matrix representation of R of T_a we have (for a simple Lie group) $\mathrm{Tr}(T_a T_b) = T(R)\delta_{ab}$, $(T_a)_{ik}(T_a)_{kj} = C_2(R)\,\delta_{ij}$ where $T(R)$ and $C_2(R)$ are Casimir coefficients. For the irreducible adjoint representation, constructed by matrices $(T_a)_{bc} = -\mathrm{i} f_{abc}$, we obtain $f_{acd} f_{bcd} = C_2(G)\,\delta_{ab} = T(G)\,\delta_{ab}$. Indicating by $(t_a)_{ij}$ the matrix generators in representation R acting on the set $\{\Phi_i\}$ the superfields are defined to transform under internal symmetry gauge transformations as

$$\Phi_i' = (\mathrm{e}^{-\mathrm{i}\Lambda_a t_a})_{ij}\Phi_j = (\mathrm{e}^{-\mathrm{i}\Lambda}\Phi)_i \tag{7.2}$$

where Λ is a Lie-algebra-valued gauge parameter superfield. For a chiral superfield to be mapped to a chiral superfield the Λ_a must themselves be chiral, $\bar{\mathrm{D}}_{\dot\alpha}\Lambda_a = 0$. Likewsie $\bar\Phi_i$ is defined to transform as

$$\bar\Phi_i' = (\bar\Phi \mathrm{e}^{\mathrm{i}\bar\Lambda})_i = \bar\Phi_j(\mathrm{e}^{\mathrm{i}\bar\Lambda_a t_a})_{ji} = (\mathrm{e}^{-\mathrm{i}\bar\Lambda_a(-\bar t_a)}\bar\Phi)_i \tag{7.3}$$

where $\mathrm{D}_\alpha\bar\Lambda_a = 0$ and we remind ourselves that the $(-\bar t_a)$ generate in general an inequivalent complex representation of T_a. The kinetic energy term $\bar\Phi_i\Phi_i$ will be left invariant if $\bar\Lambda_a = \Lambda_a$ which, as shown in Chapter 4, necessarily implies that Λ_a be space–time independent, leading to a rigid or global internal symmetry.

† For the case of ordinary field theories see, for example, Bailin and Love (1986), Huang (1982) or Lopes (1981).

In order to lift the rigid internal symmetry to a local gauge symmetry we add compensating gauge superfields V_a transforming in the adjoint representation and modify the kinetic term like we did in the Abelian case and as is customarily done in ordinary field theory. We try for the new kinetic term as suggested from the Abelian case

$$\bar{\Phi} e^{gV} \Phi = \bar{\Phi}\Phi + g\bar{\Phi} V \Phi + \frac{1}{2!} g^2 \bar{\Phi} V^2 \Phi + \dots \tag{7.4}$$

where $V = V_a t_a$. Its invariance under gauge symmetry transformations (7.2), (7.3) requires that the vector superfields have the following nonlinear transformations under the non-Abelian gauge group:

$$\begin{aligned} e^{gV'} &= e^{-i\bar{\Lambda}}\, e^{gV}\, e^{i\Lambda} \\ e^{-gV'} &= e^{-i\Lambda}\, e^{-gV}\, e^{i\bar{\Lambda}}. \end{aligned} \tag{7.5}$$

From the fact that the expansion on the RHS gives rise only to commutators of the group generators which allows us to write $V' = V'_a t_a$ it follows that (7.5) holds true independent of the representation used. We note here that for the set $\{\tilde{\Phi}_i\}$ transforming according to the complex conjugate representation, $(T_a)_{ij} = (-\bar{t}_a)_{ij}$,

$$\begin{aligned} \tilde{\Phi}' = \tilde{\Phi} e^{i\Lambda} \qquad & \bar{\tilde{\Phi}}' = e^{-i\bar{\Lambda}} \bar{\tilde{\Phi}} \\ \bar{\tilde{\Phi}} e^{gV_a(-\bar{t}_a)} \tilde{\Phi} &= \tilde{\Phi} e^{-gV} \bar{\tilde{\Phi}}. \end{aligned} \tag{7.6}$$

7.2 Field strength superfield

7.2.1 Ordinary non-Abelian gauge theory

We collect here the elements of gauging an internal symmetry in ordinary non-Abelian gauge theory. We have a set, say, of scalar fields $\{\varphi_i\}$ transforming under a representation R. The kinetic term $[-\partial^l \bar{\varphi}_i \partial_l \varphi_i]$ is not invariant under local gauge transformations $\varphi' = e^{-i\lambda}\varphi$ where $\lambda(x) = \lambda_a(x) t_a$. To restore the invariance we introduce compensating gauge connection vector fields v_a^l in the theory, define the gauge covariant derivatives $\mathscr{D}_l(v) = [\partial_l + (ig/2) t_a v_{al}]$ with respect to $\{\varphi_i\}$ and modify the kinetic term to $[-(\bar{\mathscr{D}}^l \bar{\varphi}_i)(\mathscr{D}_l \varphi_i)]$ which is gauge covariant. It follows that the gauge transformation of $v_l = v_l^a T_a$ is independent of the representation and is given by

$$v_l' = e^{-i\lambda} v_l\, e^{i\lambda} + i\, e^{-i\lambda} \partial_l\, e^{i\lambda}. \tag{7.7}$$

To write the kinetic term for the gauge field we identify the corresponding gauge covariant field strength or curvature tensor $F_{lm} = F_{lm}^a T_a$ by considering the gauge covariant commutator $[\mathscr{D}_l, \mathscr{D}_m]\varphi = -F_{lm}\varphi$ where $F_{lm} =$

$\partial_l v_m - \partial_m v_l + (ig/2)[v_l, v_m]$. They evidently transform homogeneously

$$F'_{lm} = e^{-i\lambda} F_{lm}\, e^{i\lambda} \tag{7.8}$$

and give rise to the gauge covariant kinetic term $[-\frac{1}{4}F_a^{lm}F_{lm}^a]$ for the connection fields v_l. For the case of superfields SS generalisation of the gauge transformation corresponding to (7.7) is given by nonlinear transformations (7.5). The gauge transformations of component fields in the non-Abelian case is not so simple as in the Abelian case (see (7.52)).

7.2.2 *Gauge covariant derivatives for superfields*

In order to identify a candidate for the field strength superfield which contains in it the covariant curl F_{lm}^a we may try to find it from a study of gauge and SS covariant derivatives.

Consider a set of superfields $\{F_i\}$ on which the internal symmetry gauge rotations induce the transformation

$$F' = e^{-i\Lambda} F \qquad \bar{D}_{\dot{\alpha}}\Lambda_a = 0. \tag{7.9}$$

We define the gauge and SS covariant derivatives ∇_A, $A = (l, \alpha, \dot{\alpha})$, of F by requiring that under the gauge transformations

$$(\nabla_A F') = \nabla'_A F' = e^{-i\Lambda}(\nabla_A F) = e^{-i\Lambda}\, \nabla_A\, e^{i\Lambda} F' \tag{7.10}$$

or

$$\nabla'_A = e^{-i\Lambda}\, \nabla_A\, e^{i\Lambda}. \tag{7.11}$$

The operator identities $e^{-i\Lambda}\bar{D}_{\dot{\alpha}}\, e^{i\Lambda} = \bar{D}_{\dot{\alpha}}$ and $e^{+i\bar{\Lambda}} D_\alpha e^{-i\bar{\Lambda}} = D_\alpha$ suggest that a convenient choice, for example, is $\nabla_{\dot{\alpha}} = \bar{D}_{\dot{\alpha}}$. Taking into consideration (7.5) we may choose

$$\nabla_\alpha = e^{-gV} D_\alpha\, e^{gV} \tag{7.12}$$

since

$$\nabla'_\alpha = e^{-i\Lambda}\, e^{-gV}\, e^{i\bar{\Lambda}} D_\alpha\, e^{-i\bar{\Lambda}}\, e^{gV}\, e^{i\Lambda} = e^{-i\Lambda}\, \nabla_\alpha\, e^{i\Lambda}.$$

Then because of

$$\{\nabla_\alpha, \nabla_{\dot{\alpha}}\}' = e^{-i\Lambda}\{\nabla_\alpha, \nabla_{\dot{\alpha}}\} e^{i\Lambda} \tag{7.13}$$

we may define ∇_l by

$$\{\nabla_\alpha, \nabla_{\dot{\alpha}}\} = -2i\sigma^l_{\alpha\dot{\alpha}}\, \nabla_l = -2i\, \nabla_{\alpha\dot{\alpha}}. \tag{7.14}$$

The connection superfields may be defined through $\nabla_A F = (D_A - i\Gamma_A)F$. In our asymmetric representation of ∇_α, $\nabla_{\dot{\alpha}}$ we find ($\bar{D}_{\dot{\alpha}} \equiv D_{\dot{\alpha}}$)

$$\Gamma_{\dot{\alpha}} = 0$$

$$\begin{aligned} -i\Gamma_\alpha &= e^{-gV}(D_\alpha\, e^{gV}) \\ &= gD_\alpha V - \frac{g^2}{2}[V, (D_\alpha V)] + \frac{g^3}{3!}[V, [V, D_\alpha V]] + \dots \end{aligned} \tag{7.15}$$

and from

$$\{\nabla_\alpha, \nabla_{\dot\alpha}\}F = e^{-gV}D_\alpha\, e^{gV}\bar D_{\dot\alpha}F + \bar D_{\dot\alpha}(e^{-gV}D_\alpha\, e^{gV}F)$$
$$= \{D_\alpha, \bar D_{\dot\alpha}\}F - i(\bar D_{\dot\alpha}\Gamma_\alpha)F \tag{7.16}$$

we find

$$\Gamma_{\alpha\dot\alpha} = \frac{i}{2}(\bar D_{\dot\alpha}\Gamma_\alpha) \qquad \text{or} \qquad \Gamma_l = -\frac{i}{4}\bar\sigma_l^{\dot\alpha\alpha}(\bar D_{\dot\alpha}\Gamma_\alpha). \tag{7.17}$$

The gauge covariant derivatives as defined above generate the gauge covariant generalisation of SS algebra†

$$\{\nabla_\alpha, \nabla_{\dot\alpha}\} = -2i\sigma^l_{\alpha\dot\alpha}\nabla_l = -2i\nabla_{\alpha\dot\alpha} \qquad \{\nabla_\alpha, \nabla_\beta\} = \{\nabla_{\dot\alpha}, \nabla_{\dot\beta}\} = 0. \tag{7.18}$$

To take care of the gauge covariant derivatives of $\bar F$ we do the field redefinition and define $\tilde F = \bar F e^{gV}$. The field $\tilde F$ transforms like F but in the complex conjugate representation of the gauge group

$$\tilde F' = \bar F\, e^{i\bar\Lambda}\, e^{-i\bar\Lambda}\, e^{gV}\, e^{i\Lambda} = \tilde F e^{i\Lambda} = e^{-i\Lambda^a(-\bar t_a)}\tilde F. \tag{7.19}$$

Their gauge covariant derivatives may then be obtained by applying ∇_A defined above for F‡.

7.2.3 *Gauge covariant field strength superfield*

In our representation the single independent connection superfield is the spinor superfield Γ_α. Under a gauge transformation it transforms inhomogeneously

$$\Gamma'_\alpha = e^{-i\Lambda}\Gamma_\alpha\, e^{i\Lambda} + ie^{-i\Lambda}(D_\alpha\, e^{i\Lambda}). \tag{7.20}$$

For the Abelian case it reduces to§

$$\Gamma_\alpha = igD_\alpha V \qquad \text{(Abelian gauge theory)} \tag{7.21}$$

and plausible arguments led us to define in Chapter 5 the field strength spinor superfield by $W_\alpha = (1/ig)\bar D^2\Gamma_\alpha$. It is thus suggested that we define for the non-Abelian case

$$W_\alpha = \frac{1}{ig}\bar D^2\Gamma_\alpha = \frac{1}{g}\bar D^2\, e^{-gV}(D_\alpha\, e^{gV}). \tag{7.22}$$

The gauge covariance of W_α is evident from (7.20) and (3.25)

† Note that $[\nabla_l, \nabla_{\dot\alpha}] \neq 0$, $[\nabla_l, \nabla_\alpha] \neq 0$ and $D_l = \partial_l$.

‡ In supergravitation theory a symmetric representation for ∇_A may sometimes be more convenient.

§ V is in a sense a pre-potential superfield.

$$\bar{D}^2 e^{-i\Lambda}(D_\alpha e^{i\Lambda}) = e^{-i\Lambda}([\bar{D}^2, D_\alpha] e^{i\Lambda}) = 0$$
$$W'_\alpha = e^{-i\Lambda} W_\alpha e^{i\Lambda}. \tag{7.23}$$

Alternatively, we may consider the graded commutator† of ∇_A

$$[\nabla_A, \nabla_B\} = T^C_{AB} \nabla_C - iF_{AB} \tag{7.24}$$

which defines the gauge covariant torsion tensor T^C_{AB} and field strength or curvature tensor F_{AB}. Expressing the LHS of (7.24) in term of Γ_A we find

$$[\nabla_A, \nabla_B\} = [D_A, D_B\} - i[(D_A\Gamma_B) - (-1)^{ab}(D_B\Gamma_A) - i[\Gamma_A, \Gamma_B\}]. \tag{7.25}$$

In the case, $[D_A, D_B\} = t^C_{AB} D_C = t^C_{AB}\nabla_C + it^C_{AB}\Gamma_C$, we find

$$F_{AB} = D_A\Gamma_B - (-1)^{ab} D_B\Gamma_A - i[\Gamma_A, \Gamma_B\} - t^C_{AB}\Gamma_C$$
$$T_{AB}{}^C = t_{AB}{}^C \tag{7.26}$$

where $t^l_{\dot{\alpha}\alpha} = t^l_{\alpha\dot{\alpha}} = -2i\sigma^l_{\alpha\dot{\alpha}}$ is the only non-vanishing element for $N = 1$ SS algebra.

We then derive (in our representation of ∇_A)

$$F_{\dot{\alpha}\dot{\beta}} = 0 \qquad F_{\alpha\dot{\beta}} = \bar{D}_{\dot{\beta}}\Gamma_\alpha + 2i\sigma^l_{\alpha\dot{\beta}}\Gamma_l = -2i\Gamma_{\alpha\dot{\beta}} + 2i\Gamma_{\alpha\dot{\beta}} = 0$$

$$\begin{aligned} F_{\alpha\beta} &= D_\alpha\Gamma_\beta + D_\beta\Gamma_\alpha - i(\Gamma_\alpha\Gamma_\beta + \Gamma_\beta\Gamma_\alpha) \\ &= i(D_\alpha e^{-gV})(D_\beta e^{gV}) + ie^{-gV}(D_\alpha D_\beta e^{gV}) \\ &\quad + i\, e^{-gV}(D_\alpha e^{gV})\, e^{-gV}(D_\beta e^{gV}) + (\alpha \leftrightarrow \beta) \\ &= i\, e^{-gV}(D_\alpha D_\beta e^{gV}) + (\alpha \leftrightarrow \beta) = 0 \end{aligned}$$

$$F_{\dot{\alpha}l} = \bar{D}_{\dot{\alpha}}\Gamma_l \sim -\tfrac{1}{8}\varepsilon_{\dot{\alpha}\dot{\beta}}\bar{\sigma}_l^{\dot{\beta}\beta}(\bar{D}^2\Gamma_\beta) \sim \varepsilon_{\dot{\alpha}\dot{\beta}}\bar{\sigma}_l^{\dot{\beta}\beta} W_\beta. \tag{7.27}$$

The other non-vanishing components may also be expressed in terms of W_α. Thus W_α is the gauge covariant spinor superfield we are looking for.

An expansion up to quadratic terms of (7.15) gives

$$\begin{aligned} \frac{1}{ig} T_a\Gamma^a_\alpha &= D_\alpha V + \frac{g}{2}[(D_\alpha V), V] + \ldots \\ &= T_a\left(D_\alpha V_a + \frac{ig}{2} f_{cba}(D_\alpha V_c)V_b\right) + \ldots \end{aligned} \tag{7.28}$$

or

$$\frac{1}{ig}\Gamma^a_\alpha = D_\alpha V_a - \frac{ig}{2} f_{abc} V_b(D_\alpha V_c) + \ldots. \tag{7.29}$$

† $[M, N\} = MN - (-1)^{mn} NM$ where $m = 0$ or 1 according as M is a bosonic index or a fermionic index.

Then

$$W^a_\alpha = \bar{D}^2 D_\alpha V_a - \frac{ig}{2} f_{abc}\bar{D}^2 V_b(D_\alpha V_c) + \ldots . \tag{7.30}$$

The first term on the RHS has already been computed in (5.12) and contains only λ^a_α, D^a and the Abelian part $(\partial_l v^a_m - \partial_m v^a_l)$ of the covariant curl F^a_{lm}. It is straightforward to calculate the second term if we use (4.5) and (5.6) to define the component fields. We find

$$-\frac{1}{4} W^a_\alpha = -i\lambda^a_\alpha(y) + \theta_\beta\left(\delta^\beta_\alpha D_a(y) + \frac{i}{2}(\sigma^l\bar{\sigma}^m)_\alpha{}^\beta F^a_{lm}(y)\right)$$
$$+ \theta^2\sigma^l_{\alpha\dot{\alpha}}\mathscr{D}^{ac}_l\bar{\lambda}^{\dot{\alpha}}_c(y) + \text{terms involving fields } C, \chi, \bar{\chi}, M, N \tag{7.31}$$

where

$$F^a_{lm} = \partial_l v^a_m - \partial_m v^a_l - \frac{g}{2} f_{abc} v^b_l v^c_m \tag{7.32}$$

is the covariant curl of v^a_l and

$$\mathscr{D}^l_{ab}(v) = \left(\partial^l + \frac{ig}{2} v^l\right)_{ab} = \left(\delta_{ab}\partial^l + \frac{g}{2} f_{abc} v^l_c\right) \tag{7.33}$$

is the ordinary gauge covariant derivative in the adjoint representation. Compared to the Abelian case (5.12) the curl of a bosonic gauge field is replaced by its covariant curl and the ordinary derivative of a spinor gauge field by the covariant derivative of a spinor gauge field λ^a, gaugino, transforming in the adjoint representation. The usual kinetic terms for v^a_l and λ^a_α thus do appear in $\mathrm{Tr}(W^\alpha W_\alpha)$.

7.3 Supersymmetric generalised non-Abelian gauge invariant action

7.3.1 SS action for non-Abelian gauge superfield

Following arguments similar to those of §5.4 for the Abelian case we may write

$$I_V = \frac{1}{64g^2}\left(\int d^6 s W^\alpha_a W^a_\alpha + \int d^6\bar{s}\bar{W}^a_{\dot{\alpha}}\bar{W}^{\dot{\alpha}}_a\right)$$

$$= \frac{tr}{64g^2 T(R)}\left(\int d^6 s W^\alpha W_\alpha + \int d^6\bar{s}\bar{W}_{\dot{\alpha}}\bar{W}^{\dot{\alpha}}\right)$$

$$= \int d^4x[-\tfrac{1}{4}F^a_{lm}F^{lm}_a - i\lambda^a\sigma^l\mathscr{D}^{ac}_l\bar{\lambda}_c + \tfrac{1}{2}D^a D^a]$$

+ quadratic and higher order interaction terms multiplied by at least one of the fields C, M, N, χ, $\bar{\chi}$. (7.34)

An expression in terms of V^a may be derived easily. From

$$\int d^6 s W^\alpha_a W^a_\alpha = -\frac{1}{g^2}\int d^6 s(\bar{D}^2\Gamma^\alpha_a)(\bar{D}^2\Gamma^a_\alpha)$$

$$= \frac{4}{g^2}\int d^8 z \Gamma^\alpha_a(\bar{D}^2\Gamma^a_\alpha) = -\frac{4}{g^2}\int d^8 z(D^\alpha V_a)\bar{D}^2(D_\alpha V^a) + \ldots \tag{7.35}$$

on discarding surface terms and making use of $D^\alpha\bar{D}^2 D_\alpha = \bar{D}_{\dot{\alpha}}D^2\bar{D}^{\dot{\alpha}}$ we find

$$I_V = \frac{1}{8}\int d^8 z V^a(D^\alpha\bar{D}^2 D_\alpha V^a) + \text{interaction terms}$$

$$= -\int d^8 z V^a \Box P_T V_a + \ldots . \tag{7.36}$$

apart from a suitable gauge-fixing action along with the induced ghost term to be discussed in Chapter 9.

7.3.2 *Supersymmetric Yang–Mills theory with general interaction*†

The general action of $N=1$ superfields $\{\Phi_i\}$ carrying a representation R of a non-Abelian group G interacting with $N=1$ gauge superfields V^a which is invariant under supersymmetry and non-Abelian gauge transformations (7.2), (7.3), (7.5) is given by

$$I = \int d^8 z\bar{\Phi}_i(e^{gV})_{ij}\Phi_j + \int d^6 s W(\Phi) + \int d^6\bar{s}\bar{W}(\bar{\Phi}) + I_V \tag{7.37}$$

apart from gauge-fixing and ghost terms which are necessary when we quantise the theory. Furthermore the superpotential must satisfy the gauge symmetry constraint

$$\frac{\delta W}{\delta\Phi_i}(t_a)_{ij}\Phi_j = 0. \tag{7.38}$$

An N-extended supersymmetric gauge theory, for example, $N=2, 4$ SS Yang–Mills theories, may sometimes be rewritten in the general form (7.37) involving only $N=1$ superfields but with additional global internal ($R-$) symmetries.

7.3.3 *Wess–Zumino gauge*

In order to appreciate the spectrum of physical fields we will express (7.37) in terms of component fields in the gauge defined by

† Ferrara and Zumino (1974); Salam and Strathdee (1974b).

$C^a = \chi^a = \bar{\chi}^a = M^a = N^a = 0$. The possibility of realising such a gauge even in the non-Abelian case follows from the expression given below for infinitesimal gauge transformations:

$$\delta V = \frac{\mathrm{i}}{g} L_{gV/2}[(\Lambda + \bar{\Lambda}) + (\coth L_{gV/2})(\Lambda - \bar{\Lambda})] + \mathrm{O}(\Lambda^2, \Lambda\bar{\Lambda}, \bar{\Lambda}^2)$$

$$= \frac{\mathrm{i}}{g}(\Lambda - \bar{\Lambda}) + \frac{\mathrm{i}}{2}[V, \Lambda + \bar{\Lambda}] + \frac{\mathrm{i}}{12} g[V, [V, \Lambda - \bar{\Lambda}]] + \ldots. \tag{7.39}$$

This result may be derived on using the following identity for infinitesimal a, b. If

$$\mathrm{e}^a \mathrm{e}^x \mathrm{e}^b = \mathrm{e}^{x'} \tag{7.40}$$

then

$$x' = x + L_{x/2}(b - a) + (L_{x/2} \coth L_{x/2})(b + a) + \mathrm{O}(b^2, ab, a^2) \tag{7.41}$$

where

$$L_{x/2} a = \left[\frac{x}{2}, a\right] \qquad \mathrm{e}^{L_{x/2}} a = \mathrm{e}^{x/2} a\, \mathrm{e}^{-x/2}$$
$$L \coth L = 1 + \frac{L^2}{3} - \frac{L^4}{45} + \frac{2L^6}{945} - \ldots. \tag{7.42}$$

Consider first the Abelian case. We may write $(g = 1)$

$$V = V_{\mathrm{WZ}} + \mathrm{i}(\bar{\Lambda}_0 - \Lambda_0) \tag{7.43}$$

where V_{WZ} is given in (5.29) and

$$\Lambda_0 = U\left(\frac{\mathrm{i}C}{2} - \theta\chi - \frac{1}{2}(M + \mathrm{i}N)\theta^2\right). \tag{7.44}$$

The kinetic term takes the form $\bar{\Psi} \exp(V_{\mathrm{WZ}})\Psi$ where $\Psi = \exp(-\mathrm{i}\Lambda_0)\Phi$ is chiral but not a scalar superfield. In the Wess–Zumino gauge the SS transformations do not leave the gauge-fixing conditions $C = \chi = \bar{\chi} = M = N = 0$ invariant. In fact, $\delta_{\mathrm{S}} C = 0$, $\delta_{\mathrm{S}} \chi_\alpha = -\mathrm{i}(\sigma^l \bar{\xi})_\alpha v_l$ and $\delta_{\mathrm{S}}(M + \mathrm{i}N) = 2\bar{\xi}\bar{\lambda}$. The gauge transformations in (5.5) also break these constraints. However, for the special choice of gauge parameters

$$\Lambda = \left(\tilde{A} = 0, \tilde{\psi} = \frac{\mathrm{i}}{\sqrt{2}}(\sigma^l \bar{\xi}) v_l, \tilde{F} = -\bar{\xi}\bar{\lambda}\right)$$

the sum of the two transformations preserves the gauge. Under these the gauge superfield transforms as

$$
\begin{aligned}
(\delta_S + \delta_G)v_l &= -i(\xi\sigma_l\bar{\lambda} - \lambda\sigma_l\bar{\xi}) \\
(\delta_S + \delta_G)\lambda &= -i\xi D - \frac{1}{2}(\sigma^m\bar{\sigma}^l)\xi F_{ml} \\
(\delta_S + \delta_G)D &= \partial_l(\xi\sigma^l\bar{\lambda} + \lambda\sigma^l\bar{\xi}) \\
(\delta_S + \delta_G)\Lambda_0 &= 0.
\end{aligned}
\tag{7.45}
$$

For the matter superfield we have

$$
\begin{aligned}
\delta_S\Psi &= e^{-i\Lambda_0}[-i\Phi\,\delta_S\Lambda_0 + \delta_S\Phi] \\
\delta_G\Psi &= e^{-i\Lambda_0}[-i\Phi\,\delta_G\Lambda_0 + \delta_G\Phi].
\end{aligned}
\tag{7.46}
$$

In the Wess–Zumino gauge and for the special choice of gauge parameters made above we obtain

$$
(\delta_S + \delta_G)\Psi = (\delta_S + \delta_G)\Phi = \delta_S\Phi - i\Lambda\Phi \tag{7.47}
$$

or

$$
\begin{aligned}
(\delta_S + \delta_G)A &= \sqrt{2}\,\xi\psi \\
(\delta_S + \delta_G)\psi &= -\sqrt{2}\,\xi F - i\sqrt{2}\,(\sigma^l\bar{\xi})\mathcal{D}_l A \\
(\delta_S + \delta_G)F &= i\sqrt{2}\,\mathcal{D}_l\psi\sigma^l\bar{\xi} + iA\bar{\xi}\bar{\lambda}
\end{aligned}
\tag{7.48}
$$

where $\mathcal{D}_l(v) = (\partial_l + (i/2)v_l)$ is the covariant derivative. The action written in the Wess–Zumino gauge is left invariant under the simultaneous transformations (7.45) and (7.48). Returning to the non-Abelian case the special class of gauge transformations with $\widetilde{A}_a = 0$ satisfy

$$
\begin{aligned}
\Lambda_a &= U[\sqrt{2}\,\theta\widetilde{\psi}_a + \theta^2\widetilde{F}_a] \\
\Lambda_a\Lambda_b &= U[-\theta^2\widetilde{\psi}_a\widetilde{\psi}_b] \\
\Lambda_a\Lambda_b\Lambda_c &= 0.
\end{aligned}
\tag{7.49}
$$

In the Wess–Zumino gauge ($V \to V^W$) we have

$$
\begin{aligned}
V_a^W &= -\theta\sigma^l\bar{\theta}v_l^a + i\theta^2\bar{\theta}\bar{\lambda}_a - i\bar{\theta}^2\theta\lambda_a + \tfrac{1}{2}\theta^2\bar{\theta}^2 D_a \\
V_a^W V_b^W &= -\tfrac{1}{2}\theta^2\bar{\theta}^2 v_l^a v_b^l \\
V_a^W V_b^W V_c^W &= 0 \qquad V_a^W V_b^W \Lambda_c = 0 \qquad \text{etc}
\end{aligned}
\tag{7.50}
$$

and the special infinitesimal gauge transformations are thus given by

$$
\delta_G V^W = \frac{i}{g}(\Lambda - \bar{\Lambda}) + \frac{i}{2}[V^W, \Lambda + \bar{\Lambda}]
$$

or

$$\delta_G V_a^W = \frac{i}{g}(\Lambda_a - \bar{\Lambda}_a) - \frac{1}{2} f_{abc} V_b^W (\Lambda_c + \bar{\Lambda}_c). \tag{7.51}$$

Written in terms of the component fields the non-vanishing variations are

$$\delta_G \chi_a = \frac{\sqrt{2}}{g} \bar{\psi}_a \qquad \delta_G \tfrac{1}{2}(M_a + iN_a) = \frac{1}{g} \tilde{F}_a$$

$$\delta_G \lambda_a = -\frac{i}{2\sqrt{2}} f_{abc} (\sigma^l \bar{\tilde{\psi}}_c) v_{lb} \qquad \delta_G \bar{\lambda}_a = -\frac{i}{2\sqrt{2}} f_{abc} (\bar{\sigma}^l \tilde{\psi}_c) v_{lb}$$

$$\delta_G D_a = -\frac{i}{\sqrt{2}} f_{abc} (\tilde{\psi}_c \lambda_b - \bar{\tilde{\psi}}_c \bar{\lambda}_b). \tag{7.52}$$

The SS transformations computed from (5.7) in the Wess–Zumino gauge then suggest that if we choose the gauge parameter functions such that

$$\tilde{\psi}_a = \frac{ig}{\sqrt{2}} (\sigma^l \bar{\xi}) v_{la} \qquad \bar{\tilde{\psi}}_a = \frac{ig}{\sqrt{2}} (\bar{\sigma}^l \xi) v_{la} \qquad \tilde{F}^a = -g \xi \lambda_a \tag{7.53}$$

$$(\delta_S + \delta_G)\{C_a, M_a, N_a, \chi_a, \bar{\chi}_a\} = 0 \tag{7.54}$$

and

$$\begin{aligned} (\delta_S + \delta_G) v_{la} &= -i(\xi \sigma_l \bar{\lambda}_a - \lambda_a \sigma_l \bar{\xi}) \\ (\delta_S + \delta_G) \lambda_{\alpha a} &= -i \xi_\alpha D_a - \tfrac{1}{2} (\sigma^m \bar{\sigma}^l)_\alpha{}^\beta \xi_\beta F^a_{ml} \\ (\delta_S + \delta_G) D_a &= \xi \sigma^l \mathscr{D}_l^{ab}(v) \bar{\lambda}_b + \mathrm{CC}. \end{aligned} \tag{7.55}$$

The transformation in (7.55) are the same as those of (5.7) in the Wess–Zumino gauge except that the ordinary curl is now replaced by the covariant curl F^a_{ml} and the ordinary derivative by the covariant derivative $\mathscr{D}_l$, and the transformations are consequently nonlinear.

For the scalar superfield the gauge transformation $\delta_G \Phi = -i t_a \Lambda^a \Phi$ combined with the SS transformation give on using (7.53)

$$\begin{aligned} (\delta_G + \delta_S) A_i &= -\sqrt{2}\, \xi \psi_i \\ (\delta_G + \delta_S) \psi_i &= -i(t_a)_{ij} A_j \tilde{\psi}_a - \sqrt{2}\,[\xi F_i + i(\sigma^l \bar{\xi}) \partial_l A_i] \\ &= -\sqrt{2}\, \xi F_i - i\sqrt{2}\,(\sigma^l \bar{\xi}) \mathscr{D}_l^{ij}(v) A_j \\ (\delta_G + \delta_S) F_i &= -i(t_a)_{ij} [A_j \tilde{F}_a - \tilde{\psi}_a \psi_j] + i\sqrt{2} (\partial_l \psi_i) \sigma^l \bar{\xi} \\ &= i\sqrt{2}\,(\mathscr{D}_l^{ij}(v) \psi_j) \sigma^l \bar{\xi} + ig(\xi \lambda)_{ij} A_j. \end{aligned} \tag{7.56}$$

The ‘kinetic term’ for matter fields expressed in the Wess–Zumino gauge reads as (see (5.32) for the Abelian case)

$$\int \mathrm{d}^8 z \bar{\Phi}\, \mathrm{e}^{gV^{\mathrm{W}}} \Phi = \int \mathrm{d}^8 z (\bar{\Phi}_i \Phi_i + g \bar{\Phi}_i V^{\mathrm{W}}_{ij} \Phi_j + \tfrac{1}{2} g^2 \bar{\Phi}_i V^{\mathrm{W}^2}_{ij} \Phi_j)$$

$$= \int \mathrm{d}^4 x \Big(\bar{A} \square A - \mathrm{i} \psi \sigma^l \partial_l \bar{\psi} + \bar{F} F$$

$$- \tfrac{1}{2} g [\mathrm{i}(\partial^l \bar{A}) v_l A - \mathrm{i} \bar{A} v_L \partial^l A - \bar{\psi} \bar{\sigma}^l v_l \psi]$$

$$+ \frac{\mathrm{i}}{\sqrt{2}} g (\bar{A} \lambda \psi - \bar{\psi} \bar{\lambda} A) + \frac{1}{2} g \bar{A} (D - \tfrac{1}{2} g v^l v_l) A \Big)$$

$$= \int \mathrm{d}^4 x \Big(-(\bar{\mathscr{D}}_l \bar{A})(\mathscr{D}^l A) - \mathrm{i} \psi \sigma^l \bar{\mathscr{D}}_l \bar{\psi} + \bar{F} F$$

$$+ \frac{\mathrm{i}}{\sqrt{2}} g (\bar{A} \lambda \psi - \bar{\psi} \bar{\lambda} A) + \frac{1}{2} g \bar{A} D A \Big) \tag{7.57}$$

where, for example, $\lambda = \lambda^a t_a$, $\bar{F}F = \bar{F}_i F_i$, $\bar{A}\lambda\psi = \bar{A}_i \lambda^a (t_a)_{ij} \psi_j$, $(\bar{\mathscr{D}}_l \bar{A})_i = [\partial_l + (\mathrm{i}g/2) v_l^a (-t_a^*)]_{ij} \bar{A}_j = (\partial_l - (\mathrm{i}g/2)\bar{v}_l)_{ij} \bar{A}_j$, $(\mathscr{D}_l A)_i = (\partial_l + (\mathrm{i}g/2) v_l)_{ij} A_j$, $(\bar{v}_l \bar{A})(v^l A) = \bar{A}_j (\bar{v}_l)_{ij} (v^l)_{ik} A_k = \bar{A}_j v_{lji} v^l_{ik} A_k = \bar{A} v_l v^l A$, etc. The kinetic term (7.57) is clearly invariant under the (combined) transformations (7.55) and (7.56). The same is true for the first surviving term of I_V in (7.34). We thus require the gluon field v_l^a accompanied by the fermionic superpartner gluino field λ_a and the auxiliary field D_a all in the adjoint representation to describe a supersymmetric gauge theory.

7.3.4 *Scalar potential*

The scalar potential (see (4.27)) now gets a contribution from the gauge interaction terms. Eliminating the auxiliary field D_a using its equation of motion

$$D_a + \tfrac{1}{2} g \bar{A}_i (t_a)_{ij} A_j = 0 \tag{7.58}$$

the scalar potential now becomes

$$V = |F_i|^2 + \tfrac{1}{2} D_a D_a. \tag{7.59}$$

When the gauge group contains U(1) factors the complete scalar potential is given by

$$V = |F_i|^2 + \sum \frac{1}{2} \left(\frac{g}{2}\right)^2 (\bar{A} t_a A)^2 + \sum \frac{1}{2} (g_1 \bar{A} Y A + \varkappa)^2 \tag{7.60}$$

where the sum in the second term on the RHS is over the several groups contained in the semi-simple group and the various chiral superfield representations while in the last term the sum is over all the U(1) factors already discussed in §5.6.

7.4 *N*-extended SS Yang–Mills theory

The $N=1$ superfield formulation we have been discussing may under certain conditions describe a theory which possesses N-extended ($N>1$) supersymmetry. The $N=1$ SS is manifest in the formulation and realised linearly while the additional supersymmetries are in general realised nonlinearly. The importance of the $N=1$ formulation derives from the fact that the auxiliary fields for N-extended supermultiplets for $N>3$ are not unambiguously known†. The $N=2$ superfields are difficult to handle at the quantised level contrary to the case of $N=1$ superfields. We will discuss below $N=2,4$ extended SS Yang–Mills (SSYM) theories which are currently of much theoretical interest.

Consider the general SS action (7.37) for the case of n sets of chiral superfields $\Phi_i=\Phi_i^a T_a$, $i=1,2,\ldots,n$, where each set with fixed i, $\{\Phi_i^a\}$, transforms under the adjoint representation of G. The gauge transformations (7.2) and (7.3) in this case may alternatively be written as‡

$$\begin{aligned}\delta\Phi=\delta\Phi_a T_a=-\mathrm{i}\Lambda_c(-\mathrm{i}f_{cab}\Phi_b)T_a&=-\mathrm{i}\Lambda_c[T_c,T_b]\Phi_b\\&=-\mathrm{i}[\Lambda,\Phi]\\ \delta\bar{\Phi}=-\mathrm{i}\bar{\Lambda}_c(-\mathrm{i}f_{cba}\bar{\Phi}_b)T_a&=-\mathrm{i}[\bar{\Lambda},\bar{\Phi}].\end{aligned}\tag{7.61}$$

Their finite forms are

$$\begin{aligned}\Phi'&=\mathrm{e}^{-\mathrm{i}\Lambda}\Phi\,\mathrm{e}^{\mathrm{i}\Lambda}\\ \bar{\Phi}'&=\mathrm{e}^{-\mathrm{i}\bar{\Lambda}}\,\bar{\Phi}\,\mathrm{e}^{\mathrm{i}\bar{\Lambda}}.\end{aligned}\tag{7.62}$$

The matter part of the action for N-extended SSYM theory may then be written in the following convenient form

$$\begin{aligned}I_{\text{matter}}=\frac{1}{C_2(G)}\mathrm{Tr}\bigg[&\int \mathrm{d}^8z\,\mathrm{e}^{-gV}\bar{\Phi}_i\,\mathrm{e}^{gV}\Phi_i\\&+\frac{2\mathrm{i}\gamma}{3!}\varepsilon_{ijk}\left(\int \mathrm{d}^6s\,\Phi_i[\Phi_j,\Phi_k]+\int \mathrm{d}^6\bar{s}\,\bar{\Phi}_i[\bar{\Phi}_j,\bar{\Phi}_k]\right)\\&+\int \mathrm{d}^6s\,m_{ij}\Phi_i\Phi_j+\int \mathrm{d}^6\bar{s}\,m_{ij}\bar{\Phi}_i\bar{\Phi}_j\bigg].\end{aligned}\tag{7.63}$$

Here $m_{ij}=m_{ji}$, ε_{ijk}, the completely antisymmetric tensor, are group invariant parameters and $i=1,2,\ldots,n=(N-1)$. The mass term is invariant under $N=1$ SS but they break N-extended SS and are called soft SS breaking terms. In the absence of matter, $n=0$, we obtain $N=1$ SSYM theory. The $N=2$ SSYM theory is obtained for $n=1$, $\gamma=0$, $m_{ij}=0$ while $N=4$

† Rivelles and Taylor (1981), Taylor (1982).
‡ In adjoint representation $(-T^{a*})=T^a=(-\mathrm{i}f_{abc})$ and hence $\bar{\mathscr{D}}_l(v)=\mathscr{D}_l(v)$.

SSYM theory with $n=3$, $m_{ij}=0$ allows for a single coupling constant γ. It is straightforward to show that the superpotential may be rewritten as

$$W(\Phi_i{}^a) = m_{ij}\,\delta_{ab}\Phi_i{}^a\Phi_j{}^b - \frac{\gamma}{3}\,g_{(ia),\,(jb),\,(kc)}\Phi_i{}^a\Phi_j{}^b\Phi_k{}^c \tag{7.64}$$

where

$$g_{(ia),\,(jb),\,(kc)} = \varepsilon_{ijk} f_{abc} \tag{7.65}$$

is completely symmetric under the interchange of the pair of indices. The kinetic term in (7.63) may also be rewritten as $\bar{\Phi}_i{}^a(\mathrm{e}^{gV})_{ab}\Phi_i{}^b$. The action in terms of component fields in the Wess–Zumino gauge follows from (7.57).

For $N=2$ SSYM theory involving chiral superfields Φ^a transforming in the adjoint representation we find from (7.57)

$$\begin{aligned} I_{\text{matter}} = \int \mathrm{d}^4x\Big(&-(\mathscr{D}^l\bar{A}_a)(\mathscr{D}_lA_a) - \mathrm{i}\psi_a\sigma^l\mathscr{D}_l\bar{\psi}_a + \bar{F}_aF_a \\ &+\frac{g}{\sqrt{2}}\,f_{abc}(\bar{A}_a\lambda_c\psi_b - \bar{\psi}_a\bar{\lambda}_cA_b) - \frac{\mathrm{i}g}{2}\,f_{abc}\bar{A}_aD_cA_b\Big) \end{aligned} \tag{7.66}$$

$$I = I_{\text{matter}} + I_V = \int \mathrm{d}^4x\mathscr{L}$$

where I_V is given in (7.34). The 2-spinors ψ_a and $\bar{\lambda}_a$ may be combined to define the Dirac 4-spinor

$$\begin{pmatrix} \psi_{a\alpha} \\ \bar{\lambda}_a^{\dot{\alpha}} \end{pmatrix}.$$

Hence we obtain in the four-component notation

$$\begin{aligned} \mathscr{L} = &-\frac{1}{4}\,F^a_{lm}F_a^{lm} - (\mathscr{D}_l\bar{A}_a)(\mathscr{D}^lA_a) + \mathrm{i}\bar{\psi}_a\gamma^l\mathscr{D}_{lab}\psi_b \\ &+\frac{g}{\sqrt{2}}\,f_{abc}\bar{\psi}_c\left(\frac{\mathrm{i}}{2}(A_a - \bar{A}_a) - \frac{1}{2}(A_a + \bar{A}_a)\gamma_5\right)\psi_b \\ &-\frac{\mathrm{i}g}{2}\,f_{abc}\bar{A}^aA^bD^c + \frac{1}{2}D_aD_a + \bar{F}_aF_a. \end{aligned} \tag{7.67}$$

The equations of motion of the auxiliary field are $F_a=0$, $D_c = -(g/2)\bar{A}_a(-\mathrm{i}f_{cab})A_b = (\mathrm{i}/2)gf_{abc}\bar{A}_aA_b$ and $V=\frac{1}{2}D_aD_a \geqslant 0$.

For the case of the SU(2) gauge group, for example, the SS ground state is obtained for $\varepsilon_{abc}A_aA_b=0$. This is solved by $A_a = (\text{const})\times n_a$ where n_a is a fixed unit vector. The SU(2) gauge symmetry is thus spontaneously broken to U(1) gauge symmetry corresponding to rotations about the preferred direction n_a.

In the case of $N=4$ SSYM theory if $m_{ij} = m\,\delta_{ij}$ we clearly have a global SU(3) symmetry with respect to the internal index $i=1,2,3$. In the case where $m_{ij} = 0$ we find an additional global U(1) symmetry which is an R-symmetry discussed in §§4.2 and 5.6 with $R(V) = 0$, $R(\Phi) = \frac{2}{3}$. The manifest global internal symmetry of the $N = 4$ SSYM theory action written in the $N = 1$ superfield formulation is thus SU(3) × U(1). The action written in terms of component fields may, however, be shown to possess an SU(4) global symmetry. We may collect the four Weyl spinors ψ_1, ψ_2, ψ_3 and λ in the representation {4} of SU(4). In fact, $R(\psi_1) + R(\psi_2) + R(\psi_3) + R(\lambda) = 0$ shows that the $U(1)_R$ generator (which is traceless) belongs to SU(4) if $(\psi_1, \psi_2, \psi_3, \lambda)^{\mathrm{T}}$ transforms according to {4}. The complex spin-zero fields A_i, $\bar{A}_i$ which give rise to three scalar plus three pseudoscalar real spin-0 fields may be assigned to the real representation {6} of SU(4) ~ O(6) through a self-conjugate tensor field $\Phi_{ij} = \frac{1}{2}\varepsilon_{ijkl}\bar{\Phi}^{kl}$ where $i = 1, 2, 3, 4$.

Problems

7.1 Show that (7.5) is valid in any representation of the group generators.

7.2 Compute $[\nabla_l, \nabla_\alpha]$, $[\nabla_l, \nabla_{\dot{\alpha}}]$, $[\nabla_\alpha, \nabla_{\dot{\alpha}}]$.

7.3 Compute the other elements of F_{AB} and show that they can all be expressed in terms of W_α.

7.4 Show that $\bar{\mathrm{D}}^2 V^b(\mathrm{D}_\alpha V^c)| = -2(M^b - \mathrm{i}N^b)\chi^c_\alpha + 4\mathrm{i}C^b\lambda^c_\alpha - 2\mathrm{i}\bar{\chi}^b_{\dot{\alpha}}(v^{c\dot{\alpha}}_\alpha - \mathrm{i}\partial_\alpha{}^{\dot{\alpha}}C^c)$

7.5 Calculate the g^2 term in W_α.

7.6 Find the first few self-interaction terms in (7.36).

7.7 For $\Lambda^a = (0, \tilde{\psi}^a, \tilde{F}^a)$ show that $\Lambda^a\Lambda^b\Lambda^c = 0$ and

$$\begin{aligned} \mathrm{e}^{\mathrm{i}\Lambda} &= U[1 + \mathrm{i}\sqrt{2}\,\theta\tilde{\psi} + \theta^2(\tfrac{1}{2}\tilde{\psi}\tilde{\psi} + \mathrm{i}\tilde{F})] \\ &= 1 + \mathrm{i}[\sqrt{2}\theta\tilde{\psi} + (\mathrm{i}/\sqrt{2})\theta^2\bar{\theta}\bar{\sigma}^l\partial_l\tilde{\psi} + \theta^2(\tfrac{1}{2}\tilde{\psi}\tilde{\psi} + \mathrm{i}\tilde{F})] \\ \mathrm{e}^{-\mathrm{i}\Lambda} &= 1 - \mathrm{i}[\sqrt{2}\bar{\theta}\bar{\tilde{\psi}} + (\mathrm{i}/\sqrt{2})\bar{\theta}^2\theta\sigma^l\partial_l\bar{\tilde{\psi}} + \bar{\theta}^2(\tfrac{1}{2}\bar{\tilde{\psi}}\bar{\tilde{\psi}} - \mathrm{i}\bar{\tilde{F}})]. \end{aligned}$$

7.8 Compute Γ_α and W_α in the Wess–Zumino gauge as well as $\delta_{\mathrm{G}}V$ for the special choice of Λ with $A = 0$.

7.9 Write the $N = 4$ SSYM theory in component fields in the Wess–Zumino gauge and verify the SU(4) internal symmetry invariance. Find the other SS transformations in this case as well as for $N = 2$ SSYM theory.

7.10 Show that

$$\Phi^a_i(\mathrm{e}^{gV})_{ab}\Phi^b_i = \frac{1}{C_2(G)}\,\mathrm{Tr}\,\mathrm{e}^{-gV}\bar{\Phi}_i\,\mathrm{e}^{gV}\Phi_i.$$

8

INTEGRATION OVER GRASSMANN VARIABLES. SUPERMATRICES

8.1 Introduction: Berezin integral†

The action is usually written as a Lagrangian density, which is a functional of the fields and their derivatives, integrated over the whole space–time volume. The result is translation invariant if the surface terms vanish. We would like analogously to write the action as a functional of the superfields for a supersymmetric theory. The Lagrangian density is now a superfield and the integration is performed over the x_m as well as over the anticommuting Grassmann variables θ and $\bar{\theta}$. The invariance of the action under the supertranslations (3.7) generated by Q, $\bar{Q}$ would imply a supersymmetric action. The formulation of a field theory with supersymmetry takes a very compact form in terms of superfields in contrast to its expression in terms of the component fields. The calculation of radiative corrections in higher loops now becomes manageable in particular due to the rules of integration by parts over full superspace (§8.2) and the superfield propagators (Chapter 9) which already incorporate in them in a certain sense the mutual cancellation of the divergences arising from the bosonic and fermionic component fields. It is also possible to implement the powerful tool of the superfield path integral over the superfields. We proceed then to discuss the notion of integration over an anticommuting Grassmann variable.

Consider the case of one $\theta, \theta^2 = 0$. We require the integration operator $\int \mathrm{d}\theta$ (or $\int_\theta$) to be translation invariant and linear in analogy to the ordinary integral

$$\int_{-\infty}^{\infty} f(x)\,\mathrm{d}x = \int_{-\infty}^{\infty} f(x+a)\,\mathrm{d}x. \tag{8.1}$$

Thus

$$\int \mathrm{d}\theta\, f(\theta) = \int \mathrm{d}\theta\, f(\theta+\alpha). \tag{8.2}$$

Now for any general $f(\theta)$ we have the expansion

$$f(\theta) = \alpha + \theta\beta \tag{8.3}$$

† Berezen (1966), de Witt (1984).

so that $\int d\theta[a+\theta\beta] = \int d\theta[a+\theta\beta+\alpha\beta]$ leads to

$$\int d\theta\ \alpha\beta = 0. \tag{8.4}$$

We may then define $\int d\theta = 0$ and to have a non-trivial operator define $\int d\theta\ \theta = 1$, the normalisation being arbitrary. We find

$$\int d\theta\ f(\theta) = \int d\theta[a+\theta\beta] = \beta = \frac{\partial}{\partial\theta} f(\theta) \tag{8.5}$$

i.e. $\int d\theta$ is really a derivative ($\int d\theta \equiv \partial/\partial\theta$).

From

$$\frac{\partial}{\partial\theta_\alpha}\frac{\partial}{\partial\theta_\beta} = -\frac{\partial}{\partial\theta_\beta}\frac{\partial}{\partial\theta_\alpha} \tag{8.6}$$

it follows that

$$\int d\theta \frac{\partial}{\partial\theta} f = \frac{\partial}{\partial\theta}\frac{\partial}{\partial\theta} f = 0 \tag{8.7}$$

which implies the validity of the rule of integration by parts given by

$$\int d\theta\ f\frac{\partial}{\partial\theta} g = \mp \int d\theta\left(\frac{\partial f}{\partial\theta}\right) g. \tag{8.8}$$

Here the upper sign holds if f is 'even' or bosonic while the lower holds if f is 'odd' or fermionic. This property justifies the introduction of the symbol $\int d\theta$ ('odd' operator).

From $\int d\theta\ \theta = 1$ it follows that the dimensional of $d\theta$, $[d\theta]$, is the inverse of $[\theta]$, contrary to the case of dx. The inverse of the Jacobian appears on a change of variable in θ, for example, if $\theta' = a\theta$ then $d\theta' = (1/a)\, d\theta$ so that $\int d\theta'\theta' = \int d\theta\ \theta = 1$. We may define also a delta function over a Grassmann variable by

$$\delta(\theta-\theta') = (\theta-\theta') \tag{8.9}$$

which vanishes for $\theta=\theta'$. Also

$$\delta(\theta-\theta')\delta(\theta-\theta') = (\theta-\theta')(\theta-\theta') = -\theta'\theta - \theta\theta' = 0. \tag{8.10}$$

In fact

$$\begin{aligned}\int d\theta\ (\theta-\theta')[a+\theta\beta] &= \int d\theta[\theta a - \theta' a - \theta'\theta\beta] \\ &= a - \int d\theta\ \theta'\theta\beta = a + \int d\theta\ \theta\theta'\beta = a + \theta'\beta\end{aligned} \tag{8.11}$$

or

$$\int d\theta\ \delta(\theta-\theta') f(\theta) = f(\theta'). \tag{8.12}$$

For n variables $\theta_1, \theta_2, \ldots, \theta_n$ we define

$$\int d^n\theta(\theta_1\theta_2 \ldots \theta_n) = \int d\theta_n \, d\theta_{(n-1)} \ldots d\theta_1\theta_1\theta_2 \ldots \theta_n = 1$$

$$\int d^n\theta(\text{Polynomial in } \theta \text{ of order } (n-1)) = 0$$

$$\int d^n\theta[af(\theta) + bg(\theta)] = a\int d^n\theta f(\theta) + b\int d^n\theta g(\theta) \qquad (a, b \text{ even}). \tag{8.13}$$

We also have

$$d^n\theta' = \left[\det \frac{\partial\theta'}{\partial\theta}\right]^{-1} d^n\theta$$

$$\delta^n(\theta - \theta') = \prod_{j=1}^{n} (\theta_j - \theta_j')$$

$$\int d^n\theta\delta^n(\theta - \theta')f(\theta) = f(\theta'). \tag{8.14}$$

For the case of variables $x_l, \theta_\alpha, \bar{\theta}_{\dot{\alpha}}$ it is convenient to choose the definition of Berezin integrals as follows. We define

$$\begin{aligned} \int d^2\theta\theta^2 &\equiv \int d^2\theta(\theta^\alpha\theta_\alpha) = 1 \\ \int d^2\bar{\theta}\bar{\theta}^2 &\equiv \int d^2\bar{\theta}(\bar{\theta}_{\dot{\alpha}}\bar{\theta}^{\dot{\alpha}}) = 1 \end{aligned} \tag{8.15}$$

with all other integrals vanishing. From $\theta^\alpha\theta_\alpha = -2\theta^1\theta^2$ and

$$\frac{1}{2}\int (d\theta^1)(d\theta^2)(-2\theta^1\theta^2) = +\int d\theta^1\theta^1 \cdot \int d\theta^2\theta^2 = 1$$

we find

$$d^2\theta = \tfrac{1}{2}\, d\theta^1 \, d\theta^2 = -\tfrac{1}{4}(d\theta^\alpha)(d\theta_\alpha) = -\tfrac{1}{4}(d\theta)(d\theta).$$

Similarly, from $\bar{\theta}_{\dot{\alpha}}\bar{\theta}^{\dot{\alpha}} = \bar{\theta}_{\dot{1}}\bar{\theta}^{\dot{1}} + \bar{\theta}_{\dot{2}}\bar{\theta}^{\dot{2}} = 2\bar{\theta}^{\dot{1}}\bar{\theta}^{\dot{2}}$

and

$$\int \left(-\frac{1}{2}\, d\bar{\theta}^{\dot{1}} \, d\bar{\theta}^{\dot{2}}\right)\bar{\theta}_{\dot{\alpha}}\bar{\theta}^{\dot{\alpha}} = +\int d\bar{\theta}^{\dot{1}}\bar{\theta}^{\dot{1}} \int d\bar{\theta}^{\dot{2}}\bar{\theta}^{\dot{2}} = 1$$

we get

$$d^2\bar{\theta} = -\tfrac{1}{2}(d\bar{\theta}^{\dot{1}})(d\bar{\theta}^{\dot{2}}) = -\tfrac{1}{4}\, d\bar{\theta}_{\dot{\alpha}} \, d\bar{\theta}^{\dot{\alpha}} = -\tfrac{1}{4}\,(d\bar{\theta})(d\bar{\theta}).$$

Also

$$\int d^2\theta = \frac{1}{2}\frac{\partial}{\partial\theta^1}\frac{\partial}{\partial\theta^2} = \frac{1}{4}\left(\frac{\partial}{\partial\theta_1}\frac{\partial}{\partial\theta^1} + \frac{\partial}{\partial\theta_2}\frac{\partial}{\partial\theta^2}\right) = \tfrac{1}{4}\,\partial^\alpha\partial_\alpha = \tfrac{1}{4}\,\partial^2$$

and analogously

$$\int d^2\bar\theta = \tfrac{1}{4}\,\bar\partial_{\dot\alpha}\bar\partial^{\dot\alpha} = \tfrac{1}{4}\,\bar\partial^2.$$

They are consistent with the normalisation condition above, since $\partial^2\theta^2 = 4$, $\bar\partial^2\bar\theta^2 = 4$ etc. We may also define delta functions with respect to the measures $d^2\theta$ and $d^2\bar\theta$ as follows

$$\begin{aligned}\delta^2(\theta-\theta') &= (\theta-\theta')^2 = (\theta-\theta')^\alpha(\theta-\theta')_\alpha = -2(\theta^1-\theta^{1\prime})(\theta^2-\theta'^2)\\ \delta^2(\bar\theta-\bar\theta') &= (\bar\theta-\bar\theta')^2 = (\bar\theta_{\dot\alpha}-\bar\theta_{\dot\alpha}{}')(\bar\theta^{\dot\alpha}-\bar\theta^{\dot\alpha}) = 2(\bar\theta^{\dot 1}-\theta^{\dot 1\prime})(\bar\theta^{\dot 2}-\bar\theta^{\dot 2\prime}).\end{aligned} \tag{8.16}$$

For example,

$$\int d^2\theta\delta^2(\theta-\theta')f(\theta^1,\theta^2,\bar\theta^{\dot 1},\bar\theta^{\dot 2}) = +\int d\theta^2 d\theta^1(\theta^1-\theta^{1\prime})(\theta^2-\theta^{2\prime})f$$

$$= \int d\theta^2\, d\theta^1(\theta^1-\theta^{1\prime})(\theta^2-\theta^{2\prime})[f(0,\theta^2,\bar\theta^{\dot 1},\bar\theta^{\dot 2}) + \theta^1 f_1(\theta^2,\bar\theta^{\dot 1},\bar\theta^{\dot 2})]$$

$$= \int d\theta^2[(\theta^2-\theta^{2\prime})f(0,\theta^2,\bar\theta^{\dot 1},\bar\theta^{\dot 2}) - \theta^{1\prime}(\theta^2-\theta^{2\prime})f_1(\theta^2,\bar\theta^{\dot 1},\bar\theta^{\dot 2})]$$

$$= \int d\theta^2(\theta^2-\theta^{2\prime})[f(0,\theta^2,\bar\theta^{\dot 1},\bar\theta^{\dot 2}) + \theta^{1\prime} f_1(\theta^2,\bar\theta^{\dot 1},\bar\theta^{\dot 2})]$$

$$= \int d\theta^2(\theta^2-\theta^{2\prime})f(\theta^{1\prime},\theta^2,\bar\theta^{\dot 1},\bar\theta^{\dot 2}) \to f(\theta^{1\prime},\theta^{2\prime},\bar\theta^{\dot 1},\bar\theta^{\dot 2}) \text{ etc.}$$

For $\theta=\theta'$, $\bar\theta=\bar\theta'$, $\delta^2(\theta-\theta')\to 0$, $\delta^2(\bar\theta-\bar\theta')\to 0$ and $\delta^2(\theta-\theta')\delta^2(\theta-\theta') = [(\theta-\theta')^2]^2 = 0$, $[\delta^2(\bar\theta-\bar\theta')]^2 = 0$. We also introduce a delta function with respect to the measure $d^4\theta \equiv d^2\theta d^2\bar\theta$ by $\delta^4(\theta-\theta') = \delta^2(\theta-\theta')\delta^2(\bar\theta-\bar\theta') = (\theta-\theta')^2(\bar\theta-\bar\theta')^2$.

Under a space–time volume integral if we adopt the convention of *dropping total divergences* (i.e. *surface integrals*) we find

$$\begin{aligned}\int d^6 s &= \int d^4x\,(\tfrac{1}{4}\partial^2) = \int d^4x\,(-\tfrac{1}{4}D^2)\\ \int d^6\bar s &= \int d^4x\,(\tfrac{1}{4}\bar\partial^2) = \int d^4x\,(-\tfrac{1}{4}\bar D^2)\\ \int d^8 z &\equiv \int d^4x\,\frac{D^2\bar D^2}{16} = \int d^4x\,\frac{\bar D^2 D^2}{16} = \int d^4x\,\frac{D^\alpha\bar D^2 D_\alpha}{16} \text{ etc}\end{aligned} \tag{8.17}$$

where $d^6s = d^4x d^2\theta$, $d^6\bar{s} = d^4x d^2\bar{\theta}$, $d^8z = d^4x d^2\theta d^2\bar{\theta} = d^4x d^4\theta$. It follows that $\int d^8z\, D_\alpha(\;) = \int d^8z\, \bar{D}_{\dot{\alpha}}(\;) = 0$.

8.2 Rules of integration by parts

The rules belows are a consequence of the relationship

$$\int d^8z\, D_\alpha(FG) = \int d^8z\, \bar{D}_{\dot{\alpha}}(FG) = 0 \tag{8.18}$$

and the derivative property of linear D operators. We have

$$\begin{aligned} \int d^8z F D_\alpha G &= \mp \int d^8z\, (D_\alpha F)G \\ \int d^8z F \bar{D}_{\dot{\alpha}} G &= \mp \int d^8z \int z\, (\bar{D}_{\dot{\alpha}} F)G \end{aligned} \tag{8.19}$$

according as F is an 'even' or 'odd' Grassmann function. We may derive other rules†. For example, from

$$D^2(FG) = (D^2F)G \pm 2(D^\alpha F)(D_\alpha G) + F(D^2G) \tag{8.20}$$

and $\int d^8z D^2(\;\;) = 0$ it follows that for F 'even' or 'odd'

$$\begin{aligned} \int d^8z(D^2F)G &= \mp 2 \int d^8z(D^\alpha F)(D_\alpha G) - \int d^8z F(D^2G) \\ &= 2 \int d^8z F(D^2G) - \int d^8z F(D^2G) \\ &= \int d^8z F(D^2G). \end{aligned} \tag{8.21}$$

Similarly for F 'even' or 'odd'

$$\int d^8z(\bar{D}^2F)G = \int d^8z F(\bar{D}^2G). \tag{8.22}$$

These could as well be derived using the rules above. We derive similarly

$$\begin{aligned} \int d^8z F D^\alpha \bar{D}^2 D_\alpha G &= \mp \int d^8z(\bar{D}^2 D^\alpha F) D_\alpha G \\ &= - \int d^8z(D_\alpha \bar{D}^2 D^\alpha F)G = \int d^8z(D^\alpha \bar{D}^2 D_\alpha F)G \end{aligned}$$

† Upper sign if for F 'even' (bosonic), lower for F 'odd' (fermionic).

$$\int d^8z F(D^2\bar{D}^2 G) = \int d^8z(\bar{D}^2 D^2 F)G$$

$$\int d^8z F(\bar{D}^2 D^2 G) = \int d^8z(D^2\bar{D}^2 F)G \tag{8.23}$$

$$\int d^8z F[(D^2\bar{D}^2 + \bar{D}^2 D^2)G] = \int d^8z[(D^2\bar{D}^2 + \bar{D}^2 D^2)F]G.$$

It is worth remembering that the rules above require integration over the whole of superspace. We may in general convert an integral over part of superspace to an integral over the whole of superspace. For example, for Φ and J chiral superfields

$$\int d^6s J\Phi = \int d^6s J\frac{\bar{D}^2 D^2}{16\Box}\Phi = \int d^6s\left(\frac{\bar{D}^2 D^2}{16\Box}J\right)\Phi$$

$$= \int d^6s(-\tfrac{1}{4}\bar{D}^2)\left[J\left(-\frac{1}{4}\frac{D^2}{\Box}\right)\Phi\right] = \int d^8z J\left(-\frac{D^2}{4\Box}\right)\Phi$$

$$\int d^6\bar{s}\bar{J}\bar{\Phi} = \int d^8z\bar{J}\left(-\frac{\bar{D}^2}{4\Box}\right)\bar{\Phi} \tag{8.24}$$

$$\int d^6s\Phi^3 = \int d^8z\Phi\left(-\frac{D^2}{4\Box}\right)\Phi^2 = \int d^8z\Phi^2\left(-\frac{D^2}{4\Box}\right)\Phi \qquad \text{etc.}$$

The following rules are very useful when the operators act on delta functions. We have†

$$\partial_l\delta^4(x-x') = -\partial_l'\delta^4(x-x')$$

$$\Box\delta^4(x-x') = \Box'\delta^4(x-x')$$

$$f(x)\partial^n\delta^4(x-x') = (-1)^n(\partial^n f(x))\delta^4(x-x')$$

$$\partial_\alpha\delta^2(\theta-\theta') = -\partial_\alpha'\delta^2(\theta-\theta') = 2(\theta_\alpha-\theta_\alpha') \tag{8.25}$$

$$\partial_{\dot\alpha}\delta^2(\bar\theta-\bar\theta') = -\partial'_{\dot\alpha}\delta^2(\bar\theta-\bar\theta') = -2(\bar\theta_{\dot\alpha}-\bar\theta_{\dot\alpha}')$$

$$\partial^2\delta^2(\theta-\theta') = \partial'^2\delta^2(\theta-\theta') = 4$$

$$\bar\partial^2\delta^2(\bar\theta-\bar\theta') = \bar\partial'^2\delta^2(\bar\theta-\bar\theta') = 4.$$

It follows then that ‡

$$D_\alpha\delta^8(z-z') = -D_\alpha'\delta^8(z-z')$$

$$\bar{D}_{\dot\alpha}\delta^8(z-z') = -\bar{D}'_{\dot\alpha}\delta^8(z-z')$$

$$D^2\delta^8(z-z') = D'^2\delta^8(z-z') \tag{8.26}$$

$$\bar{D}^2\delta^2(z-z') = \bar{D}'^2\delta^8(z-z').$$

† $\delta^4(x-x') = [1/(2\pi)^4]\int d^4k\exp[ik(x-x')]$.

‡ We make use of $f(\bar\theta)(\bar\theta-\bar\theta')^2 = f(\bar\theta')(\bar\theta-\bar\theta')^2$.

We note that these relations hold even when $\delta^8(z-z')$ is replaced by $\delta^4(\theta-\theta')\exp[ik(x-x')]$. We recall (see problem 3.3) that ($\not{\partial}=\sigma^l\partial_l$)

$$\begin{aligned}
D_1^2(\theta_1-\theta_2)^2 &= -4\exp[-i(\theta_1-\theta_2)\not{\partial}_1\bar{\theta}_1] \\
\bar{D}_1^2(\bar{\theta}_1-\bar{\theta}_2)^2 &= -4\exp[i\theta_1\not{\partial}_1(\bar{\theta}_1-\bar{\theta}_2)] \\
\bar{D}_1^2D_1^2\delta^4(\theta_{12}) &= 16\exp[i(\theta_1\not{\partial}_1\bar{\theta}_1+\theta_2\not{\partial}_1\bar{\theta}_2-2\theta_1\not{\partial}_1\bar{\theta}_2)] \\
D_1^2\bar{D}_1^2\delta^4(\theta_{12}) &= 16\exp[-i(\theta_1\not{\partial}_1\bar{\theta}_1+\theta_2\not{\partial}_1\bar{\theta}_2-2\theta_2\not{\partial}_1\bar{\theta}_1)]
\end{aligned} \tag{8.27}$$

and $\delta^4(\theta_{12})=\delta^4(\theta_1-\theta_2)$ and note also

$$\begin{aligned}
D_1^2\delta^4(\theta_{12})\exp[ik(x_1-x_2)] &= D_2^2\delta^4(\theta_{12})\exp[ik(x_1-x_2)] \\
\bar{D}_1^2D_1^2\delta^4(\theta_{12})\exp[ik(x_1-x_2)] &= \bar{D}_1^2D_2^2\delta^4(\theta_{12})\exp[ik(x_1-x_2)] \\
&= D_2^2\bar{D}_1^2\delta^4(\theta_{12})\exp[ik(x_1-x_2)] \\
&= D_2^2\bar{D}_2^2\delta^4(\theta_{12})\exp[ik(x_1-x_2)]
\end{aligned} \tag{8.28}$$

etc, and where we use for $z_i \neq z_j$

$$\begin{aligned}
\{D_{i\alpha},\bar{D}_{j\dot{\beta}}\} &= 0 \\
[D_i^2,\bar{D}_j^2] &= 0 \qquad \text{etc.}
\end{aligned} \tag{8.29}$$

We may also introduce operators in momentum space by replacing $\partial_l \to ik_l$

$$\begin{aligned}
D_{1\alpha}(k) &= \frac{\partial}{\partial\theta_1^\alpha}-\sigma^l_{\alpha\dot{\alpha}}\bar{\theta}_1^{\dot{\alpha}}k_l \\
\bar{D}_{1\dot{\alpha}}(k) &= -\frac{\partial}{\partial\bar{\theta}_1{}^{\dot{\alpha}}}+\theta_1^\alpha\sigma^l_{\alpha\dot{\alpha}}k_l=[D_{1\alpha}(-k)]^*
\end{aligned} \tag{8.30}$$

etc. Here $D_{1\alpha}(k)\equiv D_\alpha(k,\theta_1,\bar{\theta}_1)$ and $\bar{D}_{1\dot{\alpha}}(k)\equiv\bar{D}_{\dot{\alpha}}(k,\theta_1,\bar{\theta}_1)$. From the relations in configuration space we derive the *transfer rule*

$$D_\alpha(k,\theta_1,\bar{\theta}_1)\delta^4(\theta_{12})=-D_\alpha(-k,\theta_2,\bar{\theta}_2)\,\delta^4(\theta_{12}) \tag{8.31}$$

or

$$D_{1\alpha}(k)\delta^4(\theta_{12})=-D_{2\alpha}(-k)\delta^4(\theta_{12}). \tag{8.32}$$

Similarly, we obtain

$$\bar{D}_{1\dot{\alpha}}(k)\delta^4(\theta_{12})=-\bar{D}_{2\dot{\alpha}}(-k)\delta^4(\theta_{12}). \tag{8.33}$$

Other relations in momentum space then follow. For example,

$$\begin{aligned}
D_1^2(k)\delta^4(\theta_{12}) &= D_2^2(-k)\delta^4(\theta_{12}) \\
\bar{D}_1^2(k)D_1^2(k)\delta^4(\theta_{12}) &= \bar{D}_1^2(k)D_2^2(-k)\delta^4(\theta_{12}) \\
&= D_2^2(-k)\bar{D}_1^2(k)\delta^4(\theta_{12})=D_2^2(-k)\bar{D}_2^2(-k)\delta^4(\theta_{12}).
\end{aligned} \tag{8.34}$$

The commutation relations become†

$$\begin{aligned}\{\mathrm{D}_\alpha(p,\theta,\bar\theta),\bar{\mathrm{D}}_{\dot\beta}(q,\theta,\bar\theta)\} &= \sigma^l_{\alpha\dot\beta}(q+p)_l\\ \{\mathrm{D}_\alpha(p,\theta^i,\bar\theta^i),\mathrm{D}_\beta(q,\theta^j,\bar\theta^j)\} &= \{\bar{\mathrm{D}}_{\dot\alpha}(p,\theta^i,\bar\theta^i),\bar{\mathrm{D}}_\beta(q,\theta^j\bar\theta^j)\}=0\end{aligned} \tag{8.35}$$

and on using the identity $[A,BC]=\{A,B\}C-B\{A,C\}$ we find $(\mathrm{D}_\alpha(p)=\mathrm{D}_\alpha(p,\theta,\bar\theta))$

$$\begin{aligned}[\bar{\mathrm{D}}_{\dot\alpha}(p),\mathrm{D}^2(q)] &= -2(p+q)_l\mathrm{D}^\beta(q)\sigma^l_{\beta\dot\alpha}\\ [\mathrm{D}_\alpha(p),\bar{\mathrm{D}}^2(q)] &= 2(p+q)_l\sigma^l_{\alpha\dot\beta}\bar{\mathrm{D}}^{\dot\beta}(q)\\ [\mathrm{D}^2(p),\bar{\mathrm{D}}^2(q)] &= 2(p+q)_l\sigma^l_{\alpha\dot\beta}[\mathrm{D}^\alpha(p),\bar{\mathrm{D}}^{\dot\beta}(q)]\\ &= -4(p+q)^2-4(p+q)_{\alpha\dot\beta}\bar{\mathrm{D}}^{\dot\beta}(q)\bar{\mathrm{D}}^\alpha(p)\\ &= 4(p+q)^2+4(p+q)_{\alpha\dot\beta}\mathrm{D}^\alpha(p)\bar{\mathrm{D}}^{\dot\beta}(q)\end{aligned} \tag{8.36}$$

$$\begin{aligned}\mathrm{D}^2(p)\bar{\mathrm{D}}^2(q)\mathrm{D}^2(p) &= -4(p+q)^2\mathrm{D}^2(p)\\ \bar{\mathrm{D}}^2(q)\mathrm{D}^2(p)\bar{\mathrm{D}}^2(q) &= -4(p+q)^2\bar{\mathrm{D}}^2(q)\end{aligned} \tag{8.36}$$

We note also the operator identities ($\not k=\sigma^l k_l$)

$$\begin{aligned}\mathrm{e}^{-\theta\not k\bar\theta}\mathrm{D}_\alpha(k,\theta,\bar\theta)\,\mathrm{e}^{\theta\not k\bar\theta} &= \partial_\alpha\\ \mathrm{e}^{\theta\not k\bar\theta}\bar{\mathrm{D}}_{\dot\alpha}(k,\theta,\bar\theta)\,\mathrm{e}^{-\theta\not k\bar\theta} &= -\bar\partial_{\dot\alpha}\\ \mathrm{e}^{-\theta\not k\bar\theta}\bar{\mathrm{D}}_{\dot\alpha}(k,\theta,\bar\theta)\,\mathrm{e}^{\theta\not k\bar\theta} &= -\bar\partial_{\dot\alpha}+2(\theta\not k)_{\dot\alpha}\\ \mathrm{e}^{\theta\not k\bar\theta}\mathrm{D}_\alpha(k,\theta,\bar\theta)\,\mathrm{e}^{-\theta\not k\bar\theta} &= \partial_\alpha-2(\not k\bar\theta)_\alpha.\end{aligned} \tag{8.37}$$

We derive from (8.27) that

$$\begin{aligned}\delta^4(\theta_{12})\mathrm{D}_1^2\bar{\mathrm{D}}_1^2\delta^8(z_{12}) &= \delta^4(\theta_{12})\bar{\mathrm{D}}_1^2\mathrm{D}_1^2\delta^8(z_{12})=\delta^4(\theta_{12})\mathrm{D}_1^\alpha\bar{\mathrm{D}}_1^2\mathrm{D}_{1\alpha}\delta^8(z_{12})\\ &= \delta^4(\theta_{12})\bar{\mathrm{D}}_{1\dot\alpha}\mathrm{D}_1^2\bar{\mathrm{D}}_1^{\dot\alpha}\delta^8(z_{12})=16\delta^8(z_{12})\\ \delta^4(\theta_{12})\mathrm{D}_1^\alpha\bar{\mathrm{D}}_1^2\mathrm{D}_{1\beta}\delta^8(z_{12}) &= 8\delta^\alpha_\beta\delta^8(z_{12})\\ \delta^4(\theta_{12})\bar{\mathrm{D}}_{1\dot\alpha}\mathrm{D}_1^2\bar{\mathrm{D}}_1^{\dot\beta}\delta^8(z_{12}) &= 8\delta^{\dot\beta}_{\dot\alpha}\delta^8(z_{12}).\end{aligned} \tag{8.38}$$

In order to have a non-vanishing result we must have an even number of D's and $\bar{\mathrm{D}}$'s. For example, we find

$$\begin{aligned}\delta^4(\theta_{12})\mathrm{D}_{1\alpha}\bar{\mathrm{D}}_{1\dot\beta}\delta^8(z_{12}) &= \delta^4(\theta_{12})\mathrm{D}_1^2\bar{\mathrm{D}}_{1\dot\alpha}\delta^8(z_{12})\\ &= \delta^4(\theta_{12})\mathrm{D}_1^2\delta^8(z_{12})=\delta^4(\theta_{12})\mathrm{D}_{1\alpha}\bar{\mathrm{D}}_{1\dot\beta}\bar{\mathrm{D}}_{1\dot\sigma}\delta^8(z_{12})=0.\end{aligned} \tag{8.39}$$

8.3 Variational derivative of chiral superfield‡

The constraint $\bar{\mathrm{D}}_{\dot\alpha}\Phi=0$ leads to $\Phi(z)=\mathrm{e}^{\mathrm{i}\theta\partial\bar\theta}\Psi(x,\theta)=\Psi(y,\theta)$. It is

† See Chapter 3. Note that $\bar{\mathrm{D}}_{\dot\alpha}=(\mathrm{D}_\alpha)^*$, $\mathrm{D}_\alpha(k)^*=\bar{\mathrm{D}}_{\dot\alpha}(-k)$.

‡ See also Salam and Strathdee (1978).

suggested then that we write

$$\frac{\delta\Phi(z)}{\delta\Phi(z')} = \frac{\delta\Psi(y,\theta)}{\delta\Psi(y',\theta')} = \delta^4(y-y')\delta^2(\theta-\theta') \tag{8.40}$$

since $\Psi(y,\theta)$ does not satisfy any (differential) constraint. Now

$$\begin{aligned}\delta^4(y-y')\delta^2(\theta-\theta') &= \mathrm{e}^{\mathrm{i}\theta\bar{\partial}\bar{\theta}}\,\mathrm{e}^{\mathrm{i}\theta'\bar{\partial}'\bar{\theta}'}\delta^4(x-x')\delta^2(\theta-\theta')\\ &= \mathrm{e}^{\mathrm{i}\theta\bar{\partial}\bar{\theta}}\,\mathrm{e}^{-\mathrm{i}\theta\bar{\partial}\bar{\theta}'}\delta^2(\theta-\theta')\delta^4(x-x')\\ &= \delta^2(\theta-\theta')\left(-\frac{\bar{\mathrm{D}}^2}{4}\right)\delta^2(\bar{\theta}-\bar{\theta}')\delta^4(x-x'). \end{aligned} \tag{8.41}$$

Thus

$$\frac{\delta}{\delta\Phi(x',\theta',\bar{\theta}')}\Phi(x,\theta,\bar{\theta}) = -\tfrac{1}{4}\,\bar{\mathrm{D}}^2\delta^8(z-z') = -\tfrac{1}{4}\bar{\mathrm{D}}'^2\delta^8(z-z'). \tag{8.42}$$

The right-hand side is sometimes called a chiral delta function. Similarly,

$$\frac{\delta}{\delta\bar{\Phi}(z')}\bar{\Phi}(z) = -\tfrac{1}{4}\,\mathrm{D}^2\delta^8(z-z') = -\tfrac{1}{4}\,\mathrm{D}'^2\delta^8(z-z'). \tag{8.43}$$

It is clear that with these expressions $\delta\Phi\,(\delta\bar{\Phi})$ is also chiral (anti-chiral). We note that

$$\begin{aligned}\frac{\delta}{\delta\Phi(z)}\int \mathrm{d}^8z'\,\Phi(z')F(z') &= -\tfrac{1}{4}\bar{\mathrm{D}}^2\int \mathrm{d}^8z'\,\delta^8(z-z')F(z')\\ &= -\tfrac{1}{4}\,\bar{\mathrm{D}}^2F(z).\end{aligned} \tag{8.44}$$

8.4 Supermatrices, Superdeterminant, Supertrace

An $(r,s)\times(r,s)$ supermatrix is a square matrix $\mathbf{M}$ with the block diagonal form

$$\mathbf{M} = \begin{pmatrix}\mathbf{A} & \mathbf{X}\\ \mathbf{Y} & \mathbf{B}\end{pmatrix} \tag{8.45}$$

where $\mathbf{A}$ is $r\times r$, $\mathbf{B}$ is $s\times s$ and they contain 'even' (Grassmann) elements while $\mathbf{X}$, $\mathbf{Y}$ contain 'odd' elements. An example is the supermatrix $\mathrm{e}_A{}^M$ in (3.55). Due to the graded algebra in the superspace we must modify the definitions of the trace and the determinant such that the usual properties of these quantities are preserved. The trace must be required to satisfy

$$\mathrm{Str}\,\mathbf{M}_1\mathbf{M}_2 = \mathrm{Str}\,\mathbf{M}_2\mathbf{M}_1. \tag{8.46}$$

The supertrace is defined by†

$$\mathrm{Str}\,\mathbf{M} = \mathrm{Tr}\,\mathbf{A} - \mathrm{Tr}\,\mathbf{B} \tag{8.47}$$

† Arnowitt *et al.* (1975).

In fact

$$\begin{aligned}
\mathrm{Str}(\mathbf{M}_1\mathbf{M}_2) &= \mathrm{Tr}(\mathbf{A}_1\mathbf{A}_2 + \mathbf{X}_1\mathbf{Y}_2) - \mathrm{Tr}(\mathbf{Y}_1\mathbf{X}_2 + \mathbf{B}_1\mathbf{B}_2) \\
&= \mathrm{Tr}(\mathbf{A}_2\mathbf{A}_1 - \mathbf{Y}_2\mathbf{X}_1) - \mathrm{Tr}(-\mathbf{X}_2\mathbf{Y}_1 + \mathbf{B}_2\mathbf{B}_1) \\
&= \mathrm{Tr}(\mathbf{A}_2\mathbf{A}_1 + \mathbf{X}_2\mathbf{Y}_1) - \mathrm{Tr}(\mathbf{Y}_2\mathbf{X}_1 + \mathbf{B}_2\mathbf{B}_1) \\
&= \mathrm{Str}(\mathbf{M}_2\mathbf{M}_1). \qquad (8.48)
\end{aligned}$$

The superdeterminant or Berezinian is defined by

$$\mathrm{Sdet}\ \mathbf{M} = \exp(\mathrm{Str}\ \ln \mathbf{M}) \qquad (8.49)$$

so as to ensure the property Sdet $(\mathbf{M}_1\mathbf{M}_2) = (\mathrm{Sdet}\ \mathbf{M}_1)(\mathrm{Sdet}\ \mathbf{M}_2)$. For $\mathbf{M} = \exp \mathbf{C}$ we find Sdet $\mathbf{M} = \exp(\mathrm{Str}\ \mathbf{C})$. For a unimodular super or graded matrix Sdet $\mathbf{M} = 1$ and Str $\mathbf{C} = 0$. In explicit form†

$$\begin{aligned}
\mathrm{Sdet}\ \mathbf{M} &= \frac{\det(\mathbf{A} - \mathbf{X}\mathbf{B}^{-1}\mathbf{Y})}{\det \mathbf{B}} \\
&= \frac{\det \mathbf{A}}{\det(\mathbf{B} - \mathbf{Y}\mathbf{A}^{-1}\mathbf{X})} \\
&= (\det \mathbf{A})(\det \mathbf{D}) \qquad (8.50)
\end{aligned}$$

where $\mathbf{D}$ is the Fermi–Fermi part of $\mathbf{M}^{-1}$ while $\mathbf{A}$ is the Bose–Bose part of $\mathbf{M}$. In path integrals the Faddeev–Popov determinant is indeed this superdeterminant. We also note that

$$\delta(\mathrm{Sdet}\ \mathbf{M}) = \delta \exp(\mathrm{Str}\ \ln \mathbf{M}) = (\mathrm{Sdet}\ \mathbf{M}) \cdot \mathrm{Str}(\mathbf{M}^{-1}\delta\mathbf{M}) \qquad (8.51)$$

corresponding to the ordinary case $\delta \ln \det \mathbf{A} = \mathrm{Tr}(\mathbf{A}^{-1}\delta\mathbf{A})$.

Problems

8.1 Prove (8.14), (8.35), (8.36), (8.37) and (8.38).

8.2 Construct the super-Lorentz transformation matrix with respect to the flat superspace metric given in (3.41).

8.3 Show that in analogy to the conformal group defined over Minkowski space coordinates we may define over the coordinate superspace spanned by z^M with flat metric given in (3.41) the generators for super-translations, Lorentz transformations, dilatations and special conformal transformations. Show that the algebra closes under the graded commutator defined in (1.76). See *Lett. Nuovo Cim.* **17** (1976) 357 and **14** (1975) 471.

† See appendix to van Nieuwenhuizen (1981) and de Witt (1984).

8.4 Show that the functional matrix for the transformation (3.7)

$$\frac{\partial(x'^l, \theta'^\alpha, \bar{\theta}'_{\dot{\alpha}})}{\partial(x^m, \theta^\beta, \bar{\theta}_{\dot{\beta}})}$$

may be written as exp(**J**) where the non-vanishing elements of supermatrix **J** are $J^l_\beta = \mathrm{i}(\sigma^l \bar{\xi})_\beta$, $J^{l\dot{\beta}} = \mathrm{i}(\bar{\sigma}^l \xi)^{\dot{\beta}}$ and consequently it carries unit (S−) determinant. The integration measure over full superspace is thus invariant under translations (3.7) in superspace.

8.5 Demonstrate Sdet($\mathbf{M}_1\mathbf{M}_2$) = Sdet $\mathbf{M}_1$ · Sdet $\mathbf{M}_2$.
Hint:

$$(\exp \mathbf{A})(\exp \mathbf{B}) = \exp\left(\sum_{n=1}^{\infty} \frac{1}{n!} \mathbf{C}_n(\mathbf{A}, \mathbf{B})\right)$$

where $\mathbf{C}_1 = \mathbf{A} + \mathbf{B}$ and the other coefficients may be written as multi-commutators of **A** and **B**, viz, $\mathbf{C}_2 = [\mathbf{A}, \mathbf{B}]$, $\mathbf{C}_3 = \frac{1}{2}[[\mathbf{A}, \mathbf{B}], \mathbf{B}] + \frac{1}{2}[\mathbf{A}, [\mathbf{A}, \mathbf{B}]]$, $\mathbf{C}_4 = [[\mathbf{A}, [\mathbf{A}, \mathbf{B}]], \mathbf{B}]$ etc (Baker–Campbell–Haussdorff formula).

8.6 Prove the following result:

$$\delta^4(\theta - \theta')[\mathrm{D}_\alpha \bar{\mathrm{D}}_{\dot{\alpha}} \mathrm{D}^2 \bar{\mathrm{D}}^2 \delta^4(\theta - \theta')]\ \mathrm{e}^{\mathrm{i}k(x-x')} = 32\sigma^l_{\alpha\dot{\alpha}} k_l\, \mathrm{e}^{\mathrm{i}k(x-x')} \delta^4(\theta - \theta').$$

9

SUPERFIELD PROPAGATORS

9.1 Introduction

From the contents of previous chapters we found that superfields are very convenient for constructing representations of SS algebra and SS invariant actions. We will now develop the rules for quantising supersymmetric theories in terms of superfields directly. A superfield Feynman diagram contains many component-field diagrams which give rise to miraculous cancellation of divergences.

9.2 Gauge superfield propagator

The gauge superfield action I_V in (7.34) is invariant under the generalised gauge transformations (7.5). In order to define propagators (or Green functions) for a superfield V^a we must fix the gauge by adding to the action a gauge-fixing term along with the corresponding induced ghost superfield term in the action. This can be done supersymmetrically by adding a term

$$I_{GF} = -\tfrac{1}{8}\xi \int d^8z \bar{\mathscr{F}}^a \mathscr{F}^a \tag{9.1}$$

to the action, where

$$\mathscr{F}^a = D^2 V^a \qquad D_\alpha \bar{\mathscr{F}}^a = 0 \tag{9.2}$$

and ξ is the gauge parameter. We find with the normalisation chosen in (9.1) that

$$\begin{aligned} I_{GF} &= -\tfrac{1}{8}\xi \int d^8z (\bar{D}^2 V_a)(D^2 V_a) \\ &= -\tfrac{1}{16}\xi \int d^8z V_a (D^2\bar{D}^2 + \bar{D}^2 D^2) V_a \\ &= -\tfrac{1}{2}\xi (\partial_l v_a^l)^2 + \ldots \end{aligned} \tag{9.3}$$

contains the usual covariant gauge-fixing term for v_l^a. In the case of spontaneous breakdown of internal gauge symmetry (but not of SS) it is possible to modify (9.2) (Ovrut and Wess 1982) by adding a term dependent on Φ_i

(but not on V) so as to avoid the ΦV mixed propagators† in the shifted theory. This is achieved, however, at the expense of introducing new interaction terms in the action as in the ordinary theory. Limiting ourself to the simplest gauge-fixing functional (9.2) the induced superghost term in the action is given by

$$I_{\text{ghost}} = \frac{1}{C_2(G)} \operatorname{Tr} \int d^8z(\bar{C}' - C')L_{gV/2}[(C+\bar{C}) + (\coth L_{gV/2})(C-\bar{C})]$$

$$= \int d^8z(\bar{C}'_a C_a + C'_a \bar{C}_a) + \text{interactions}. \tag{9.4}$$

Here C_a, C'_a are anticommuting chiral ghost superfields. Collecting together the terms quadratic in V the free action for the vector superfield is given by

$$I_0 = \int d^8z V_a A_{ab} V_b + 2\int d^8z J_a V_a. \tag{9.5}$$

Here

$$A_{ab} = [-\Box + M^2]_{ab} P_{\mathrm{T}} + [-\xi\Box + M^2]_{ab}(P_1 + P_2) \tag{9.6}$$

where for generality we have included the contribution of an induced mass term‡ when the gauge symmetry is spontaneously broken and added an external source term. The factor 2 in the vector superfield source term is necessary to ensure the correct source term for the gluon field

$$2\int d^8z J_a V_a = -\int d^4x J^l_a v^a_l + \ldots. \tag{9.7}$$

From (9.5) it follows that

$$\delta I_0 = 2\int d^8z(A_{ab}V_b + J_a)\,\delta V_a. \tag{9.8}$$

From the Euler–Lagrange equations $\delta I_0/\delta V_a(z) = 0$ and on using $\delta V_a(z)/\delta V_b(z') = \delta_{ab}\delta^8(z-z')$ we derive

$$A_{ab}V_b = -J_a. \tag{9.9}$$

The free theory Green functions are defined by

$$\begin{aligned} A_{ac}\Delta_{cb}(z,z') &= \delta_{ab}\,\delta^8(z-z') \\ V_a(z) &= -\int d^8z'\,\Delta_{ab}(z,z')\,J_b(z'). \end{aligned} \tag{9.10}$$

In view of the properties of the projection operators P_1, P_2, P_{T} discussed

† $\mathscr{F}_a = D^2[V_a + (g/2\xi)\bar{a}_i(t_a)_{ij}(1/\Box)\Phi_j]$ where $a_i = \langle A_i\rangle_0$. Note that $\langle F_i\rangle_0 = \langle\psi_i\rangle_0 = 0$.

‡ $M^2_{ab} = \frac{1}{2}g^2\bar{a}_i(t_at_b)_{ij}a_j$; $t_at_b = (1/2)\{t_a, t_b\} + (i/2)f_{abc}t_c$.

in §3.3 we may easily invert (A_{ab}) and find for the free propagators

$$\Delta_{ab}(z,z') = \left[\left(\frac{1}{-\Box + M^2}\right)_{ab} P_T + \xi^{-1}\left(\frac{1}{-\Box + M^2/\xi}\right)_{ab} P_L\right] \delta^8(z-z') \tag{9.11}$$

where $M^2_{ab} = M^2_{ba}$ is the symmetric part of M^2. For $\xi = 1$ it reduces to the simple form

$$\Delta_{ab}(z,z') = \left(\frac{1}{-\Box + m^2}\right)_{ab} \delta^8(z-z') \qquad \xi = 1. \tag{9.12}$$

In momentum space, for example, for $\xi = 1$

$$\Delta_{ab}(k;\theta,\theta') = \left(\frac{1}{k^2 + M^2}\right)_{ab} \delta^4(\theta - \theta'). \tag{9.13}$$

9.3 Chiral superfield propagator

9.3.1 Unconstrained chiral superpotential

The chiral superfield Φ satisfies a differential constraint $\bar{D}_{\dot\alpha}\Phi = 0$ ($D_\alpha\bar\Phi = 0$), similar to the case of Abelian gauge theory field strength $\partial_l\partial_m F^{lm} = 0$. It is suggested (Srivastava 1983b) that we introduce unconstrained superfield potentials S, $\bar{S}$ such that†

$$\Phi = -\tfrac{1}{4}\bar{D}^2 S \qquad \bar\Phi = -\tfrac{1}{4}D^2\bar{S}. \tag{9.14}$$

The perturbation theory also is more conveniently done in terms of the propagators of S, $\bar{S}$ rather than those of Φ, $\bar\Phi$. For example, consider the cubic interaction term in the Wess–Zumino model (see (4.22))

$$\int d^6 s\Phi^3 + \text{CC} = \int d^8 z[S(-\tfrac{1}{4}\bar{D}^2 S)(-\tfrac{1}{4}\bar{D}^2 S) + \bar{S}(-\tfrac{1}{4}D^2\bar{S})(-\tfrac{1}{4}D^2\bar{S})] \tag{9.15}$$

where we absorb a $\bar{D}^2$ or D^2 factor to achieve integration over full superspace. From the following self-explanatory diagrams (figure 9.1) with two and three internal lines it is clear that we need the (simpler) propagators of S, $\bar{S}$ rather than those of Φ, $\bar\Phi$ along with the Feynman rules for vertices following from Wick's theorem in the conventional way: a vertex with two (three) internal lines of S or $\bar{S}$ requires a factor $(-\tfrac{1}{4}\bar{D}^2)$ or $(-\tfrac{1}{4}D^2)$ respectively on one (two) of the internal propagators. The rule is evidently generalised to the case of many different interacting chiral superfields.

† For example, $S = -(1/4\Box)D^2\Phi + \bar{D}\Sigma$, $\bar{S} = -\tfrac{1}{4}\bar{D}^2\Phi + D\Sigma$.

From (9.14) it follows that the superfield (strengths) Φ, $\bar{\Phi}$ are unaltered under the following Abelian gauge transformation

$$S \to S + \bar{D}_{\dot{\alpha}}\bar{\Sigma}^{\dot{\alpha}} \qquad \bar{S} \to \bar{S} + D^{\alpha}\Sigma_{\alpha} \tag{9.16}$$

and as such the superfield action written in terms of S, $\bar{S}$ will carry this gauge invariance associated with chiral superfields. A gauge-fixing term is required to take care of it and will be found below. Consequently, a neat formulation† of the generating functional represented by a functional integral over S, $\bar{S}$ is done in a straightforward fashion.

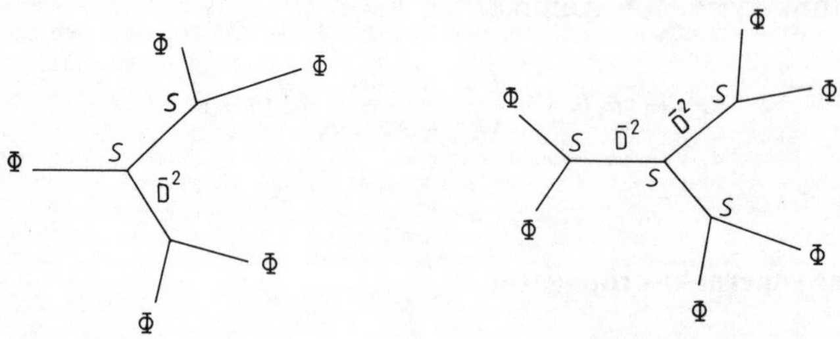

Figure 9.1

9.3.2 S, $\bar{S}$ superpropagators, Gauge-fixing term‡

Consider the action (4.23) of n interacting chiral superfields. The free action is given by

$$I_0 = \int d^8 z \bar{\Phi}_i \Phi_i + \frac{1}{2}\left(\int d^6 s m_{ij}\Phi_i\Phi_j + \int d^6 \bar{s}\bar{m}_{ij}\bar{\Phi}_i\bar{\Phi}_j\right)$$

$$= \int d^8 z\left(\bar{S}_i \Box P_1 S_i - \frac{m_{ij}}{8} S_i\bar{D}^2 S_j - \frac{\bar{m}_{ij}}{8}\bar{S}_i D^2 \bar{S}_j\right). \tag{9.17}$$

To take care of the gauge invariance (9.16) we add the gauge-fixing action

$$I_{GF} = \xi \int d^8 z \bar{S}_i \Box (1 - P_1) S_i$$

$$= -\xi \int d^8 z (\bar{D}_{\dot{\alpha}}\bar{S}_i)\left(\frac{3}{16}\bar{D}^{\dot{\alpha}}D^{\alpha} + \frac{1}{4}D^{\alpha}\bar{D}^{\dot{\alpha}}\right)(D_{\alpha}S_i) \tag{9.18}$$

so that the gauge-fixing functions are $\mathscr{F}^i_{\alpha} = D_{\alpha}S_i$, $\bar{\mathscr{F}}^i_{\dot{\alpha}} = \bar{D}_{\dot{\alpha}}\bar{S}_i$. The ghosts

† Φ and $\Phi + \bar{D}^2\Omega$ both satisfy $\bar{D}_{\dot{\alpha}}\Phi = 0$. However, it is difficult to find such a term to formulate a functional integral over Φ, $\bar{\Phi}$ contrary to that over S, $\bar{S}$.
‡ Srivastava (1984).

decouple and the complete free action becomes

$$\tilde{I}_0 = I_0 + I_{\rm GF} + I_{\rm S} = \int {\rm d}^8 z [\bar{S}_i \Box [P_1 + \xi(P_2 + P_{\rm T})] S_i$$
$$- \tfrac{1}{8}(m_{ij} S_i \bar{\rm D}^2 S_j + \bar{m}_{ij} \bar{S}_i {\rm D}^2 \bar{S}_j)] + \int {\rm d}^8 z [J_i S_i + \bar{J}_i \bar{S}_i] \tag{9.19}$$

where an external souce term has been added. Euler–Lagrange equations are obtained from

$$\frac{\delta \tilde{I}_0}{\delta S_i(z)} = 0 \qquad \frac{\delta \tilde{I}_0}{\delta \bar{S}_i(z)} = 0$$

on making use of

$$\frac{\delta S_i(z)}{\delta S_j(z')} = \frac{\delta \bar{S}_i(z)}{\delta \bar{S}_j(z')} = \delta_{ij}\, \delta^8(z - z')$$

$$\frac{\delta S_i}{\delta \bar{S}_j} = 0 \qquad \frac{\delta \bar{S}_i}{\delta S_j} = 0$$

and integrating freely by parts. We find

$$\begin{aligned} &\Box [P_2 + \xi(1 - P_2)] \bar{S}_i - \tfrac{1}{4} m_{ij} \bar{\rm D}^2 S_j = - J_i \\ &\Box [P_1 + \xi(1 - P_1)] S_i - \tfrac{1}{4} \bar{m}_{ij} {\rm D}^2 S_j = - \bar{J}_i \end{aligned} \tag{9.20}$$

or

$$A_{ab} S_b = - J_b \tag{9.21}$$

where

$$\mathbf{S} = \begin{pmatrix} S_i \\ \bar{S}_i \end{pmatrix} \quad \mathbf{J} = \begin{pmatrix} J_i \\ \bar{J}_i \end{pmatrix} \quad \mathbf{A} = \begin{pmatrix} -\tfrac{1}{4} m_{ij} \bar{\rm D}^2 & \Box [P_2 + \xi(P_1 + P_{\rm T})] \\ \Box [P_1 + \xi(P_2 + P_{\rm T})] & -\tfrac{1}{4} \bar{m}_{ij} {\rm D}^2 \end{pmatrix} \tag{9.22}$$

where $a = 1, 2, \ldots, 2n$ numbers the $2n$ superfields, say J_i, $\bar{J}_i$ in the order specified in (9.22). The free Green functions or propagators are then defined by

$$A_{ab} \Delta_{bc}(z, z') = \delta_{ac}\, \delta^8(z - z') \tag{9.23a}$$

$$S_a(z) = - \int {\rm d}^8 z' \Delta_{ab}(z, z') J_b(z') \tag{9.23b}$$

where

$$\mathbf{\Delta} = (\Delta_{ab}) = \begin{pmatrix} \Delta^{SS} & \Delta^{S\bar{S}} \\ \Delta^{\bar{S}S} & \Delta^{\bar{S}\bar{S}} \end{pmatrix} \tag{9.24}$$

and, for example, $\Delta^{S_i \bar{S}_j}(z, z')$ indicates the $S_i \bar{S}_j$ superpropagator.

It is straightforward† to invert $\mathbf{A}$ to obtain $\boldsymbol{\Delta}$ if we use the set of differential operators P_1, P_2, P_T, P_+, P_- which is closed under multiplication and which was discussed in §3.3. We may in fact rewrite

$$\mathbf{A} = -\sqrt{\Box}P_-\begin{pmatrix} m_{ij} & 0 \\ 0 & 0 \end{pmatrix} - \sqrt{\Box}P_+\begin{pmatrix} 0 & 0 \\ 0 & \bar{m}_{ij} \end{pmatrix}$$

$$+ P_1\Box\begin{pmatrix} 0 & \xi \\ 1 & 0 \end{pmatrix} + P_2\Box\begin{pmatrix} 0 & 1 \\ \xi & 0 \end{pmatrix} + P_T\Box\begin{pmatrix} 0 & 1 \\ 1 & 0 \end{pmatrix}. \tag{9.25}$$

Expressing $\boldsymbol{\Delta}$ as

$$\boldsymbol{\Delta} = \frac{1}{\sqrt{\Box}} P_+ a_+ + \frac{1}{\sqrt{\Box}} P_- \, a_- + \frac{1}{\Box} P_1 a_1 + \frac{1}{\Box} P_2 a_2 + \frac{1}{\Box} P_T a_T \tag{9.26}$$

and requiring that $\mathbf{A}\boldsymbol{\Delta} = \mathbf{I}$ we determine the a's. We obtain finally

$$\begin{aligned}
\Delta^{S\bar{S}}(z, z') &= \left(\frac{1}{\xi}\frac{(P_2 + P_T)}{\Box} + (\Box - \bar{m}m)^{-1}P_1\right) \delta^8(z - z') \\
\Delta^{\bar{S}S}(z, z') &= \left(\frac{1}{\xi}\frac{(P_1 + P_T)}{\Box} + (\Box - m\bar{m})^{-1}P_2\right) \delta^8(z - z') \\
\Delta^{SS}(z, z') &= \frac{\bar{m}}{4\Box} D^2\Delta^{\bar{S}S}(z, z') = \bar{m}(\Box - m\bar{m})^{-1} \frac{D^2}{4\Box} \delta^8(z - z') \\
\Delta^{\bar{S}\bar{S}}(z, z') &= \frac{m}{4\Box} \bar{D}^2\Delta^{S\bar{S}}(z, z') = m(\Box - \bar{m}m)^{-1} \frac{\bar{D}^2}{4\Box} \delta^8(z - z').
\end{aligned} \tag{9.27}$$

We note also that gauge parameter dependent term drops out in the following expressions

$$\begin{aligned}
\bar{D}^2\Delta^{S\bar{S}}(z, z') &= (\Box - \bar{m}m)^{-1}\bar{D}^2 \, \delta^8(z - z') \\
D'^2\Delta^{S\bar{S}}(z, z') &= (\Box - \bar{m}m)^{-1}D^2 \, \delta^8(z - z') \\
D^2\Delta^{\bar{S}S}(z, z') &= (\Box - m\bar{m})^{-1}D^2 \, \delta^8(z - z') \\
\bar{D}'^2\Delta^{\bar{S}S}(z, z') &= (\Box - m\bar{m})^{-1}\bar{D}^2 \, \delta^8(z - z')
\end{aligned} \tag{9.28}$$

if we use the properties of delta function discussed in Chapter 8.

The Φ, $\bar{\Phi}$ propagators may then be derived by acting with the appropriate D^2, $\bar{D}^2$ operators on S, $\bar{S}$ propagators. For example, using (9.28)

$$\begin{aligned}
\Delta^{\Phi\bar{\Phi}}(z, z') &= \frac{1}{16} \bar{D}^2D'^2\Delta^{S\bar{S}}(z, z') = \frac{\bar{D}^2D^2}{16} (\Box - \bar{m}m)^{-1} \delta^8(z - z') \\
&= \frac{\Box}{(\Box - \bar{m}m)} P_2 \, \delta^8(z - z') \\
\Delta^{\Phi\Phi}(z, z') &= \frac{1}{16} D^2\bar{D}'^2\Delta^{SS}(z, z') = \bar{m}(\Box - m\bar{m})^{-1} \frac{D^2}{4} \delta^8(z - z').
\end{aligned} \tag{9.29}$$

† A simpler method which is applicable even when m_{ij} are superfields is given below in §9.3.4.

From

$$\bar{\mathrm{D}}^2\,\delta^8(z-z') = -4(\theta-\theta')^2 \exp[\mathrm{i}\theta\sigma^l(\bar\theta-\bar\theta')]\,\partial_l\,\delta^4(x-x')$$
$$= -4(\theta-\theta')^2 \exp[\mathrm{i}\theta\sigma^l\bar\theta\partial_l]\,\exp[+\mathrm{i}\theta'\sigma^l\bar\theta'\partial'_l]\,\delta^4(x-x')$$
$$= -4(\theta-\theta')^2\,\delta^4(y-y') \qquad (9.30)$$

we obtain, say, for a single chiral superfield,

$$\Delta^{\Phi\Phi}(z,z') = \langle T\Phi(z)\Phi(z')\rangle_0 = \langle T\Psi(y,\theta)\Psi(y',\theta')\rangle_0$$
$$= -\bar m(\theta-\theta')^2\Delta_{\mathrm{F}}(y-y') \qquad (9.31)$$

where $\Delta_{\mathrm{F}}(x-x') = (\Box - m\bar m)^{-1}\,\delta^4(x-x')$. The propagators of the component fields then follow by making use of the expansion

$$\Psi(y,\theta)\Psi(y',\theta') = A(y)A(y') + \theta^2F(y)A(y') + \theta'^2A(y)F(y')$$
$$-2\theta^\alpha\theta'^\beta\psi_\alpha(y)\psi_\beta(y') + \ldots . \qquad (9.32)$$

The non-vanishing propagators are

$$\begin{aligned}\langle TA(x)F(x')\rangle_0 = \langle TF(x)A(x')\rangle_0 &= -\bar m\Delta_{\mathrm{F}}(x-x')\\ \langle T\psi_\alpha(x)\psi_\beta(x')\rangle_0 &= -\bar m\varepsilon_{\alpha\beta}\Delta_{\mathrm{F}}(x-x').\end{aligned} \qquad (9.33)$$

Other non-vanishing component field propagators are derived by considering $\Delta^{\Phi\bar\Phi}$ and are found to be

$$\begin{aligned}\langle TA(x)\bar A(x')\rangle_0 &= \Delta_{\mathrm{F}}(x-x')\\ \langle TF(x)\bar F(x')\rangle_0 &= \Box\Delta_{\mathrm{F}}(x-x')\\ \langle T\psi_\alpha(x)\bar\psi_{\dot\beta}(x')\rangle_0 &= -\mathrm{i}\partial_{\alpha\dot\beta}\Delta_{\mathrm{F}}(x-x').\end{aligned} \qquad (9.34)$$

The expressions in (9.33), (9.34) coincide with the propagators obtained by expressing the free action in terms of the component fields and computing the propagators directly. Historically the inverse of the procedure above was in fact followed to derive the chiral superfield propagators (Salam and Strathdee 1978, Fayet and Ferrara 1977). We remark that the gauge parameter independent part of the S, $\bar S$ propagators could as well be readily obtained from the knowledge of the Φ, $\bar\Phi$ propagators in terms of those of the component fields, the relationships in (9.29) and the known operations of D, $\bar{\mathrm{D}}$ operators on the delta function $\delta^8(z-z')$. They are, of course, not defined uniquely, as is also clear from the fact that a choice different from (9.18) for the gauge-fixing term would lead to different S, $\bar S$ propagators. The choice (9.18) leads to the 'simplified' form for the gauge parameter independent part of the propagator which is found in current literature (Grisaru *et al* 1979).

The superpropagators in momentum space are obtained from (9.27) to be

$$\Delta^{S\bar{S}}(p)=\left(-\frac{1}{\xi}\frac{P_2+P_T}{p^2}-\frac{1}{p^2+\bar{m}m}P_1\right)\delta^4(\theta-\theta')$$

$$\Delta^{\bar{S}S}(p)=\left(-\frac{1}{\xi}\frac{P_1+P_T}{p^2}-\frac{1}{p^2+m\bar{m}}P_2\right)\delta^4(\theta-\theta')$$

$$\Delta^{SS}(p)=\frac{\bar{m}}{4}\frac{1}{p^2(p^2+m\bar{m})}\mathrm{D}^2\,\delta^4(\theta-\theta') \tag{9.35}$$

$$\Delta^{\bar{S}\bar{S}}(p)=\frac{m}{4}\frac{1}{p^2(p^2+\bar{m}m)}\bar{\mathrm{D}}^2\,\delta^4(\theta-\theta')$$

where

$$\Delta(z,z')=\int\frac{\mathrm{d}^4p}{(2\pi)^4}\,\Delta(p)\,\mathrm{e}^{\mathrm{i}p(x-x')}.$$

9.3.3 *Generating functional*

The free generating functional for the chiral superfields may be written over unconstrained S, $\bar{S}$ superfields (Srivastava 1984) as

$$Z_0[J,\bar{J}]=\mathbb{N}\int[\mathrm{d}S]\,[\mathrm{d}\bar{S}]\,\exp[\mathrm{i}(I_0+I_{\mathrm{GF}})+\int\mathrm{d}^8z(JS+\bar{J}\bar{S})] \tag{9.36}$$

where $\mathbb{N}$ is the normalisation factor. We may rewrite the argument of the exponential, using (9.23*a*) and integration by parts, as

$$I_0+I_{\mathrm{GF}}+I_{\mathrm{source}}$$

$$=\int\mathrm{d}^8z\left\{\frac{1}{2}\left[\begin{pmatrix}S\\ \bar{S}\end{pmatrix}+\int\mathrm{d}^8z'\,\Delta(z,z')\begin{pmatrix}J(z')\\ \bar{J}(z')\end{pmatrix}\right]^{\mathrm{T}}A(z)\right.$$
$$\times\left[\begin{pmatrix}S\\ \bar{S}\end{pmatrix}+\int\mathrm{d}^8z'\,\Delta(z,z')\begin{pmatrix}J(z')\\ \bar{J}(z')\end{pmatrix}\right]$$
$$\left.-\frac{1}{2}\int\mathrm{d}^8z\,\mathrm{d}^8z'\begin{pmatrix}J(z)\\ \bar{J}(z)\end{pmatrix}^{\mathrm{T}}\Delta(z,z')\begin{pmatrix}J(z')\\ \bar{J}(z')\end{pmatrix}\right\} \tag{9.37}$$

where $A(z)$ is given in (9.22). The following linear substitution in the functional integral

$$\begin{pmatrix}S'(z)\\ \bar{S}'(z)\end{pmatrix}=\begin{pmatrix}S(z)\\ \bar{S}(z)\end{pmatrix}+\int\mathrm{d}^8z'\,\Delta(z,z')\begin{pmatrix}J(z')\\ \bar{J}(z')\end{pmatrix} \tag{9.38}$$

leads to the following path integral representation of the generating functional of the free Green functions

$$\frac{Z[J,\bar{J}]}{Z[0,0]} = \exp -\frac{\mathrm{i}}{2}\int \mathrm{d}^8z\,\mathrm{d}^8z'\,(J(z),\bar{J}(z))\Delta(z,z')\begin{pmatrix} J(z') \\ \bar{J}(z') \end{pmatrix} \tag{9.39}$$

where

$$Z[0,0] = \mathbb{N}\int [\mathrm{d}S'][\mathrm{d}\bar{S}']\,\exp\frac{\mathrm{i}}{2}\int \mathrm{d}^8z(S'(z),\bar{S}'(z))A\begin{pmatrix} S'(z') \\ \bar{S}'(z') \end{pmatrix}. \tag{9.40}$$

The generating functional of the interacting theory with the interaction term $\int \mathrm{d}^8z\mathscr{L}_{\mathrm{int}}[S,\bar{S}]$ is obtained to be

$$Z[J,\bar{J}] = \exp \mathrm{i}\left(\int \mathrm{d}^8z\mathscr{L}_{\mathrm{int}}\left[\frac{1}{\mathrm{i}}\frac{\delta}{\delta J},\frac{1}{\mathrm{i}}\frac{\delta}{\delta \bar{J}}\right]\right)Z_0[J,\bar{J}] \tag{9.41}$$

which forms the basis for the superfield perturbation theory. The connected Green functions are generated by $W[J,\bar{J}] = -\mathrm{i}\ln Z[J,\bar{J}]$. The Legendre transformation

$$\Gamma[S,\bar{S}] = W[J,\bar{J}] - \int \mathrm{d}^8z(JS+\bar{J}\bar{S})$$

called the effective action, generates one-particle-irreducible (OPI) S, $\bar{S}$ vertices. We remark that we are treating the unconstrained superfields on their own footing without appealing to the structure of their component fields. For the chiral superfields Φ, $\bar{\Phi}$, in view of the differential constraints it seems difficult to formulate a functional integral over them. Moreover, as emphasised earlier we do need to work with S, $\bar{S}$ propagators rather than those of Φ, $\bar{\Phi}$.

For the gauge superfield which has no differential constraint the functional integral is defined as usual on the vector superfield itself. A gauge-fixing term (see §9.2) must be added to the action along with the corresponding Faddeev–Popov superdeterminant† in the functional integral. The latter may be reinterpreted as the ghost-superfield term in the action. For the Abelian case with the gauge-fixing conditions (9.1) the ghosts decouple.

9.3.4 Alternative derivation of chiral superpropagators

In some well known methods for calculating the effective potential we are required to know the propagators when $M = (m_{ij})$ in (9.22) is a chiral superfield (see §10.4). The following straightforward method is also applicable to this general case. From (9.22), (9.23) and (9.24) we obtain ($\bar{\mathrm{D}}M = 0$ and

† See Ovrut and Wess (1982), Grisaru *et al.* (1979) and Gates *et al.* (1984).

$\mathrm{D}\bar{M}=0$)

$$\square[P_1+\xi(P_2+P_{\mathrm{T}})]\Delta^{S\bar{S}}-\tfrac{1}{4}\bar{M}\mathrm{D}^2\Delta^{\bar{S}\bar{S}}=\delta^8(z-z') \tag{9.42}$$

$$-\tfrac{1}{4}M\bar{\mathrm{D}}^2\Delta^{S\bar{S}}+\square[P_2+\xi(P_1+P_{\mathrm{T}})]\Delta^{\bar{S}\bar{S}}=0. \tag{9.43}$$

It is evident from the properties of P_1, P_2 and P_{T} that (9.43) is solved by

$$\Delta^{\bar{S}\bar{S}}=\frac{1}{4\square}M\bar{\mathrm{D}}^2\Delta^{S\bar{S}}. \tag{9.44}$$

Substituting it in (9.42) we obtain

$$\{\square[P_1+\xi(P_2+P_{\mathrm{T}})]-\bar{M}P_1M\}\Delta^{S\bar{S}}(z,z')=\delta^8(z-z'). \tag{9.45}$$

On observing that $P_1MP_1M=MP_1MP_1=MP_1M$ we may rewrite it as

$$[P_1(\square-\bar{M}P_1M)+\xi\square(P_2+P_{\mathrm{T}})]\Delta^{S\bar{S}}=\delta^8(z-z') \tag{9.46}$$

which can easily be inverted to obtain

$$\Delta^{S\bar{S}}(z,z')=\left[\frac{1}{\xi\square}(P_2+P_{\mathrm{T}})+P_1(\square-\bar{M}P_1M)^{-1}\right]\delta^8(z-z'). \tag{9.47}$$

For the cases of interest the second term may be expanded such that the θ, $\bar{\theta}$ dependence and the pole structure are made explicit allowing for easy integration over superspace variables.

Problems

9.1 Express the gauge-fixing term (9.3) in terms of the component fields.

9.2 Find the superfield propagators of the ghost superfields.

9.3 Derive the free propagators of the component fields from the Wess–Zumino Lagrangian (2.28). Construct the Φ, $\bar{\Phi}$ propagators and derive the gauge parameter independent part of the S, $\bar{S}$ propagators.

9.4 Write (9.23) explicitly for the elements in (9.24). Solve the differential equations to obtain the result in (9.27) directly.

9.5 A 'shifted' theory is obtained by shifting all the scalar fields by arbitrary constants (background fields). Consider SSQED ((5.36), Chapter 5) and find a modification of the gauge-fixing condition (9.2) such that the non-diagonal terms $\Phi_\pm V$ are removed when the auxiliary fields are not shifted. What is the modification in the general case and for the non-Abelian theory? (See footnote on page 110.)

9.6 Derive the Feynman rules for the vertices by functionally differentiating $Z[J,\bar{J}]$.

10

SUPERFIELD PERTURBATION THEORY

10.1 Introduction

The superfield Green functions to any order in perturbation theory may be constructed using Wick's theorem in the conventional way or its modern version in terms of the generating functional represented by a functional or path integral over the superfields. The computation of a superfield Feynman diagram sums up the results of a collection of several diagrams of the component formulation of an SS theory and is very economical. The cancellation of divergences of boson loops with fermion loops is automatically taken care of. The superfield path integral formulation is a powerful calculational tool. The Ward identities may be derived as usual by a change of variable and calculating the Jacobi determinant. The background field method may also be extended to the superfields. The computation of an effective potential in higher loops using superfields for SS theories, (spontaneously) broken or not, becomes manageable. The superfield tadpole and bubble (Srivastava 1985) methods of evaluating effective potentials may easily be formulated. We shall illustrate in the following sections the ease with which we can handle a superdiagram if we make judicious use of the algebra of $\mathrm{D}, \bar{\mathrm{D}}$ operators (Chapter 3) and the rules of integration by parts over the whole of superspace along with the useful identities discussed in Chapter 8. We shall consider mainly the Wess–Zumino model and refer the reader to the current literature for SS non-Abelian gauge theories, for example, $N=4$ SSYM theory which has been shown to be finite to all orders (Gates *et al.* 1983).

10.2 Wess–Zumino model

10.2.1 Tadpole diagram

The cubic interaction term in (4.34) may be rewritten as an integral over the whole of superspace

$$H_{\text{int}} = -\tfrac{1}{3}\, g_{ljk} \int \mathrm{d}^8 z \Phi_l S_j(-\tfrac{1}{4}\bar{\mathrm{D}}^2) S_k + \text{CC}. \qquad (10.1)$$

The tadpole contribution to 1-loop order is read out to be (see figure 10.1)

$$\Gamma = (-\mathrm{i})(-\tfrac{1}{3})\cdot 3 \int \mathrm{d}^8 z \bar{\Phi}_l(z) g_{ljk} [(-\tfrac{1}{4}\mathrm{D}^2)\, \Delta^{\bar{S}_j \bar{S}_k}(z, z')]_{z=z'} + \text{CC} \qquad (10.2)$$

where (– i) derives from the perturbation theory expansion and 3 is the symmetry factor at the vertex. From the expression (9.27) for the propagator and (8.38) which implies

$$D^2\bar{D}^2\delta^8(z-z')|_{\theta=\theta',\bar{\theta}=\bar{\theta}'} = 16\,\delta^4(x-x')$$

we find that

$$\Gamma \sim g_{ljk} \int d^4\theta\,[\bar{\tilde{A}}_l(0) + \sqrt{2}\,\bar{\theta}\bar{\tilde{\psi}}_l(0) + \bar{\theta}^2\bar{\tilde{F}}_l(0)]$$

$$\int d^4k(k^2)^{-1}[m(k^2+\bar{m}m)^{-1}]_{jk} = 0 \qquad (10.3)$$

where a tilde is used to indicate the Fourier transform of a component field (or a superfield). The A, ψ and F tadpole contributions as read off from (10.3) give a vanishing result and there is consequently no spontaneous SS breaking to 1-loop order in the model.

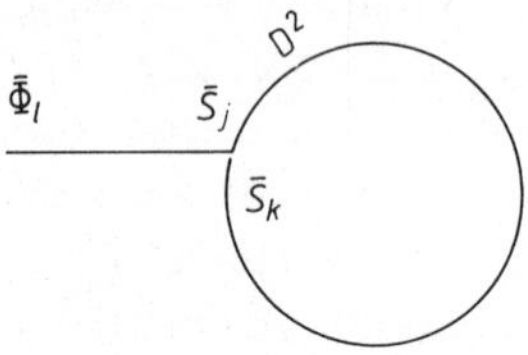

Figure 10.1

10.2.2 One-loop corrections to the $S\bar{S}$ propagator

We consider now the 1-loop corrections to the two-point Green functions of the superfields S and $\bar{S}$. For simplicity we consider a single superfield; the results are easy to extend for n interacting chiral superfields.

Consider first the corrections to the $S\bar{S}$ propagator where the 'self-energy' graphs of figure 10.2 must be considered. We shall show that (a) and (b) give a vanishing result while (c) and (d) give rise to a non-vanishing wave function renormalisation.

From the rules for the vertices found in §9.3 and indicated also in the figures we may write the contribution from figure 10.2(a) as

$$\Gamma_a \sim \int d^8z_1\, d^8z_2 \frac{d^4k\, d^4q}{k^2(k^2+|m|^2)q^2(q^2+|m|^2)}\Phi_1\Phi_2$$

$$\times\,(\bar{D}_1^2 D_1^2\delta^4(\theta_1-\theta_2)\,e^{ik(x_1-x_2)})(\bar{D}_2^2 D_2^2\delta^4(\theta_2-\theta_1)\,e^{iq(x_2-x_1)}) \qquad (10.4)$$

where $\bar{D}_1\Phi_1 = \bar{D}_2\Phi_2 = 0$ and we use the Fourier representation of the delta

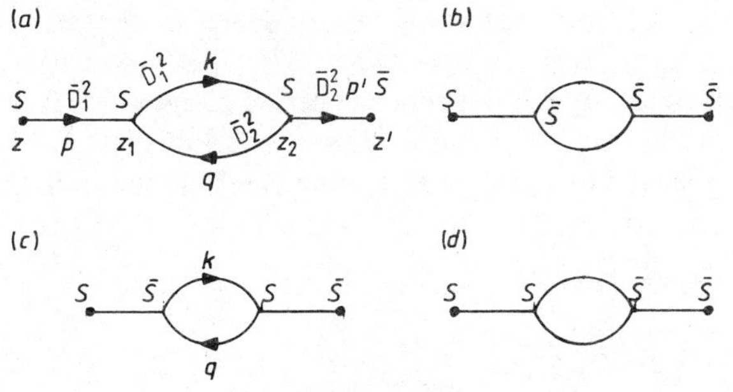

Figure 10.2

function,

$$\delta^4(x-x') = \int \frac{d^4k}{(2\pi)^4}\, e^{ik(x-x')}.$$

Since the integration is over all of superspace we may freely integrate by parts (see Chapter 8) and in view of the fact that $\bar{D}_1(\Phi_1\Phi_2) = 0$ we may rewrite the last two factors in the integrand of (10.4) as ($\delta(\theta_{12}) = \delta(\theta_1 - \theta_2)$):

$$\begin{aligned} I = {} & [D_1^2\delta^4(\theta_{12})]\,[\bar{D}_1^2\bar{D}_2^2 D_2^2\delta^4(\theta_{21})] \\ & \quad (k) \qquad\qquad\qquad (q) \end{aligned} \tag{10.5}$$

where the exponentials are understood and suppressed for the convenience of writing. The second factor may then be simplified on using

$$\bar{D}_1^2\bar{D}_2^2 D_2^2\delta^4(\theta_{21}) = \bar{D}_2^2 D_2^2\bar{D}_1^2\delta^4(\theta_{21}) = \bar{D}_2^2 D_2^2\bar{D}_2^2\delta^4(\theta_{21}) = (-16q^2)\bar{D}_2^2\delta^4(\theta_{21}) \tag{10.6}$$

where we use the identity $\bar{D}^2 D^2 \bar{D}^2 = 16\Box\bar{D}^2$ discussed in Chapter 3. We may now liberate a $\delta^4(\theta_{12})$ by performing another integration by parts ($\bar{D}_2(\Phi_1\Phi_2) = 0$) and rewrite

$$I = -16q^2\delta^4(\theta_{12})\bar{D}_2^2 D_1^2\delta^4(\theta_{12}) \to -16q^2 \cdot 16\delta^4(\theta_{12})\, e^{i(q-k)(x_2-x_1)} \tag{10.7}$$

where we use (8.27) and have restored the exponentials in the final form. In $\theta, \bar{\theta}$ space the points 1 and 2 are contracted to one point because of $\delta^4(\theta_1 - \theta_2)$. We note that $\Phi_1 = \bar{D}_1^2\,\Delta^{SS}$ is chiral in z_1 and as such all its dependence on $\bar{\theta}_1$ occurs in the operator $\exp[i\theta_1\sigma^l\bar{\theta}_1\partial_{1l}]$ which acts on the function of other variables including θ_1. In fact,

$$\Phi_1 \sim \int d^4p\, e^{ip(x-x_1)}\, e^{\theta_1 \not{p} \bar{\theta}_1} \ldots$$

and by similar arguments

$$\Phi_2 \sim \int d^4p'\, e^{ip'(x_2-x')}\, e^{-\theta_2 \not{p}' \bar{\theta}_2} \ldots$$

(where $\not{p}=\sigma^l p_l$) contains all the $\bar{\theta}_2$ dependence in the exponential. An intergration over x_1, x_2 leads to a factor $\delta^4(p-p')$ which together with $\delta^4(\theta_{12})$ cancels out all the $\bar{\theta}_1$ dependence of the integrand. Since the surviving integral on θ_1 is over the full $\theta_1, \bar{\theta}_1$ space it vanishes giving $\Gamma_a=0$.

In the case that the line ending at z_1 is an external one indicating a chiral superfield Φ_1 we obtain

$$\begin{aligned}\Phi_1(z_1) &= \exp[\mathrm{i}\theta_1\sigma^l\bar{\theta}_1\partial_{1l}]\,[A(x_1)+\sqrt{2}\,\theta_1\psi(x_1)+\theta_1^2F(x_1)] \\ &= \int\frac{\mathrm{d}^4p}{(2\pi)^4}\,\mathrm{e}^{\theta_1\not{p}\bar{\theta}_1}[\tilde{A}(-p)+\sqrt{2}\,\theta_1\tilde{\psi}(-p)+\theta_1^2\tilde{F}(-p)]\,\mathrm{e}^{-\mathrm{i}px_1} \qquad (10.8)\end{aligned}$$

where a tilde indicates the Fourier transform of the component field. It is clear that a vanishing contribution results again and the same holds even when both the lines ending at z_1 and z_2 indicate external chiral superfields. An analogous procedure shows that $\Gamma_b=0$.

We turn our attention now to figure 10.2(c). We have

$$\begin{aligned}\Gamma_c = \frac{(-\mathrm{i})^2}{2!}\left(-\frac{g}{3}\right)^2\cdot(3\cdot 3\cdot 2)\int \mathrm{d}^8z_1\,\mathrm{d}^8z_2\bar{\Phi}_1\Phi_2[-\tfrac{1}{4}\mathrm{D}_1^2\mathrm{i}\Delta^{\bar{S}S}(z_1,z_2)] \\ \times[-\tfrac{1}{4}\bar{\mathrm{D}}_2^2\mathrm{i}\Delta^{S\bar{S}}(z_2,z_1)] \qquad (10.9)\end{aligned}$$

where the factors multiplying the integral are the perturbation theory factor, coupling constant and the symmetry factor† (topological weight) respectively associated with the diagram. We also insert a factor 'i' along with each propagator as defined in Chapter 9 in order to conform to the usual practice‡ in conventional field theory. The $\bar{\Phi}_1$ and Φ_2 stand for

$$\begin{aligned}\bar{\Phi}_1 &= -\tfrac{1}{4}\mathrm{D}_1^2\mathrm{i}\Delta^{S\bar{S}}(z,z_1) = -\tfrac{1}{4}\mathrm{D}^2\int\frac{\mathrm{d}^4p}{(2\pi)^4}\,\frac{-\mathrm{i}}{p^2+|m|^2}\,\delta^4(\theta-\theta_1)\,\mathrm{e}^{\mathrm{i}p(x-x_1)} \\ \Phi_2 &= -\tfrac{1}{4}\bar{\mathrm{D}}_2^2\mathrm{i}\Delta^{S\bar{S}}(z_2,z') = -\tfrac{1}{4}\bar{\mathrm{D}}'^2\int\frac{\mathrm{d}^4p'}{(2\pi)^4}\,\frac{-\mathrm{i}}{p'^2+|m|^2}\,\delta^4(\theta_2-\theta')\,\mathrm{e}^{\mathrm{i}p'(x_2-x')}. \qquad (10.10)\end{aligned}$$

Hence

$$\Gamma_c = -g^2\cdot\frac{1}{16}\left(\frac{\mathrm{i}}{(2\pi)^4}\right)^2\int \mathrm{d}^8z_1\,\mathrm{d}^8z_2\,\frac{\mathrm{d}^4k\,\mathrm{d}^4q}{(k^2+|m|^2)(q^2+|m|^2)}\,\bar{\Phi}_1\Phi_2\cdot I \qquad (10.11)$$

† We may attach the first leg of the vertex at z_1 to z in three ways and the first leg of the vertex at z_2 to z' also in three ways. There are two ways to sew the second leg of the vertex z_1 to the vertex z_2. Thus the topological weight of the diagram is $(3\cdot 3\cdot 2)$.

‡ The Feynman propagator Δ_F satisfies $(\Box-\mu^2)\Delta_\mathrm{F}=\mathrm{i}\delta^4(x)$ and $[\varphi(\boldsymbol{x},t), \dot{\varphi}(\boldsymbol{x}',t)]=\mathrm{i}\delta^3(\boldsymbol{x}-\boldsymbol{x}')$.

where

$$I = (D_1^2\delta^4(\theta_1-\theta_2)\,e^{ik(x_1-x_2)})(\bar{D}_2^2\delta^4(\theta_2-\theta_1)\,e^{iq(x_2-x_1)}). \qquad (10.12)$$

On integrating by parts we obtain

$$I \rightarrow \delta^4(\theta_1-\theta_2)D_1^2\bar{D}_2^2\delta^4(\theta_2-\theta_1)\,e^{i(\ldots)} \rightarrow 16\,\delta^4(\theta_1-\theta_2)\,e^{i(q-k)(x_2-x_1)} \qquad (10.13)$$

which after the integrations over θ_2, x_1 and x_2 leads to

$$\Gamma_c = -g^2\frac{D^2\bar{D}'^2}{16}\left(\frac{i}{(2\pi)^4}\right)^4$$

$$\times\int d^4\theta_1\,\frac{d^4k\,d^4q\,d^4p\,d^4p'\,[(2\pi)^4]^2\delta^4(p-q-k)\delta^4(p'+q-k)}{(k^2+|m|^2)(q^2+|m|^2)(p^2+|m|^2)(p'^2|m|^2)}$$

$$\times e^{i(px-p'x')}\delta^4(\theta-\theta_1)\delta^4(\theta_1-\theta')$$

$$= -g^2\frac{D^2\bar{D}'^2}{16\Box}\int\frac{d^4p}{(2\pi)^4}\frac{-i}{(p^2+|m|^2)}\delta^4(\theta-\theta')\,e^{ip(x-x')}\frac{(-p^2)}{p^2+|m|^2}II$$

$$= g^2\int\frac{d^4p}{(2\pi)^4}\frac{-i}{p^2+|m|^2}P_1\delta^4(\theta-\theta')\frac{p^2}{p^2+|m|^2}II(p^2;|m|^2)\cdot e^{ip(x-x')}$$

(10.14)

where

$$II = i\int\frac{d^4k}{(2\pi)^4}\frac{1}{(k^2+|m|^2)[(k-p)^2+|m|^2]} \qquad (10.15)$$

is a logarithmically divergent integral.

It is clear that we may set down self-evident rules to write the contribution in momentum space directly. We will compute the contribution of figure 10.2(d) using such rules. We find

$$\Gamma_d = \frac{(-i)^2}{2!}\left(-\frac{g}{3}\right)^2\cdot(3\cdot3\cdot2)\cdot\int d^4\theta_1\,d^4\theta_2\,d^4k\,d^4p\,d^4p'\,d^4q$$

$$\times[(2\pi)^4]\,\delta^4(p+q-k)(2\pi)^4\delta^4(k-q-p')\left(\frac{i}{(2\pi)^4}\right)^4$$

$$\times\left(-\frac{1}{4}D_1^2\cdot\frac{\bar{m}}{4}\frac{1}{p^2(p^2+|m|^2)}D^2\delta^4(\theta-\theta_1)\right)$$

$$\times\left(-\frac{1}{4}\bar{D}_1^2\cdot\frac{-1}{k^2+|m|^2}\delta^4(\theta_1-\theta_2)\right)$$

$$\times\left(-\frac{1}{4}D_2^2\cdot\frac{-1}{q^2+|m|^2}\delta^4(\theta_2-\theta_1)\right)$$

(*Continued*)

$$\times\left(-\frac{1}{4}D_2^2\cdot\frac{m}{4}\frac{1}{p'^2(p'^2+|m|^2)}\bar{D}_2^2\delta^4(\theta_2-\theta')\right)$$

$$\times e^{ip(x-x')}$$

$$=-g^2\int d^4\theta_1\, d^4\theta_2\frac{d^4p}{(2\pi)^4}\frac{d^4k}{(2\pi)^4}\frac{|m|^2}{(16)^3}$$

$$\times\frac{e^{ip(x-x')}}{[p^2(p^2+|m|^2)]^2(k^2+|m|^2)(q^2+|m|^2)}I \tag{10.16}$$

where $q=(k-p)$ and

$$I=\underset{(p)}{[\bar{D}_1^2D^2\delta^4(\theta-\theta_1)]}\;\underset{(k)}{[\bar{D}_1^2\delta^4(\theta_1-\theta_2)]}\;\underset{(q)}{[D_2^2\delta^4(\theta_2-\theta_1)]}\;\underset{(p)}{[D_2^2\bar{D}_2^2\delta^4(\theta_2-\theta')]} \tag{10.17}$$

where we have indicated in I the four-momentum associated with each factor and we recall that a factor like $\bar{D}_1^2\delta^4(\theta_1-\theta_2')$ originates from a factor $\bar{D}_1^2\delta^4(\theta_1-\theta_2)\exp(ikx)$ in configuration space where integration by parts may be freely performed. Hence on an integration by parts we may liberate one factor $\delta^4(\theta_1-\theta_2)$ and use $\delta^4(\theta_1-\theta_2)\bar{D}_1^2D_2^2\delta^4(\theta_2-\theta_1)=16\delta^4(\theta_1-\theta_2)$ to obtain

$$I=16\delta^4(\theta_1-\theta_2)[\bar{D}_1^2D_1^2\delta^4(\theta-\theta_1)]\,[D_2^2\bar{D}_2^2\delta^4(\theta_2-\theta')]. \tag{10.18}$$

Integrating on θ_2 and performing another integration by parts it goes over to

$$\begin{aligned}I&=16\delta^4(\theta-\theta_1)[D_1^2\bar{D}_1^2D_1^2\bar{D}_1^2\delta^4(\theta_1-\theta')]\\&=16\delta^4(\theta-\theta_1)(-16p^2)D_1^2\bar{D}_1^2\delta^4(\theta_1-\theta')\\&=-(16)^2p^2\delta^4(\theta-\theta_1)\bar{D}'^2D'^2\delta^4(\theta_1-\theta').\end{aligned} \tag{10.19}$$

On integrating now over θ_1 we find

$$I=-(16)^2p^2\bar{D}'^2D'^2\delta^4(\theta-\theta')=+(16)^3(p^2)^2\frac{D^2\bar{D}^2}{(-16p^2)}\delta^4(\theta-\theta') \tag{10.20}$$

$$\Gamma_d=-g^2\int\frac{d^4p}{(2\pi)^4}\frac{-i}{p^2+|m|^2}P_1\delta^4(\theta-\theta')\frac{|m|^2}{p^2+|m|^2}\Pi(p^2,|m|^2)\,e^{ip(x-x')}. \tag{10.21}$$

The $S\bar{S}$ propagator corrected to 1-loop order thus reads in momentum space as

$$-i\left(\frac{1}{\xi}\frac{(P_2+P_1)}{p^2}+\frac{P_1}{(p^2+|m|^2)}(1+g^2R_1)+(g^4)\right)\delta^4(\theta-\theta') \tag{10.22}$$

where

$$R_1 = \left(\frac{p^2 - |m|^2}{p^2 + |m|^2}\right) \Pi(p^2, |m|^2). \tag{10.23}$$

The gauge parameter term gets no correction. For $m = 0$ we obtain $R_1 = \Pi(p^2)$.

10.2.3 One-loop corrections to the SS propagator

We have to consider the following non-vanishing diagrams (figure 10.3).

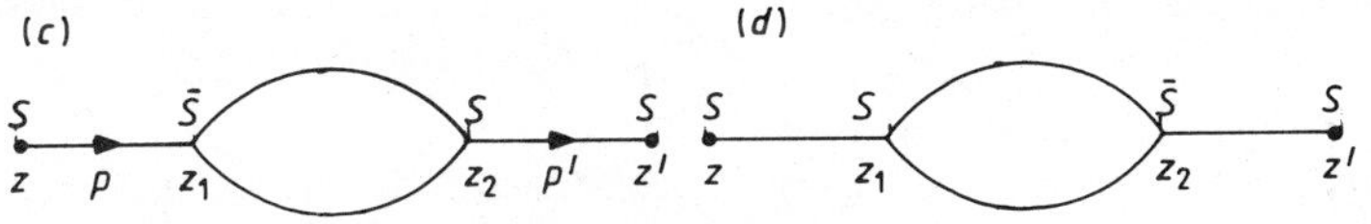

Figure 10.3

The contribution Γ_c is obtained from (10.9) on replacing Φ_2 in (10.10) by

$$\Phi_2 = -\tfrac{1}{4}\bar{D}_2^2\, i\Delta^{SS}(z_2, z') = -\tfrac{1}{4}\bar{D}_2^2 \int \frac{d^4p'}{(2\pi)^4}\, i\frac{\bar{m}}{4}\frac{1}{p'^2(p'^2 + |m|^2)} D_2^2\delta^4(\theta_2 - \theta')$$
$$\times e^{ip'(x_2 - x')}$$
$$= -\tfrac{1}{4}D'^2\bar{D}'^2\frac{i\bar{m}}{4}\int \frac{d^4p'}{(2\pi)^4}\frac{\delta^4(\theta_2 - \theta')}{p'^2(p'^2 + |m|^2)} e^{ip'(x_2 - x')}. \tag{10.24}$$

Hence

$$\Delta_c^{SS} = -g^2\left(\frac{i}{[(2\pi)^4]}\right)^4 \frac{\bar{m}}{4}(-1)$$
$$\times \int \frac{d^4k\, d^4q\, d^4p\, d^4p'\, [(2\pi)^4]^2\delta^4(q + p - k)\delta^4(q + p' - k)}{(k^2 + |m|^2)(q^2 + |m|^2)(p^2 + |m|^2)p'^2(p'^2 + |m|^2)}$$
$$\times \frac{D^2D'^2\bar{D}'^2}{16}\delta^4(\theta - \theta')\, e^{i(px - p'x')}$$
$$= -g^2\int \frac{d^4p}{(2\pi)^4}\frac{\bar{m}}{4}\frac{(-1)}{p^2(p^2 + |m|^2)} D^2\delta^4(\theta - \theta')\, e^{ip(x - x')}\frac{(-p^2)}{(p^2 + |m|^2)}$$
$$\times \int \frac{d^4k}{(2\pi)^4}\frac{1}{(k^2 + |m|^2)(q^2 + |m|^2)}$$
$$= g^2\int \frac{d^4p}{(2\pi)^4}\, i\frac{\bar{m}}{4}\frac{1}{p^2(p^2 + |m|^2)} D^2\delta^4(\theta - \theta')\frac{p^2}{(p^2 + |m|^2)}$$
$$\times \Pi(p^2, |m|^2)\, e^{ip(x - x')}. \tag{10.25}$$

Figure 10.3(d) gives the same contribution, $\Gamma_d = \Gamma_c$, and the propagator corrected to 1-loop order is

$$\left(\mathrm{i}\,\frac{\bar{m}}{4}\,\frac{1}{p^2(p^2+|m|^2)}\,\mathrm{D}^2(1+g^2R_2)+\mathrm{O}(g^4)\right)\delta^4(\theta-\theta') \qquad (10.26)$$

where

$$R_2 = 2\,\frac{p^2}{(p^2+|m|^2)}\,\Pi(p^2,|m|^2) = R_1 + \Pi. \qquad (10.27)$$

The relationship

$$\Delta^{SS}(z,z') = \frac{\bar{m}}{4\Box}\,\mathrm{D}^2\,\Delta^{\bar{S}S}(z,z')$$

which is true at the tree level, is violated by the radiative corrections.

We note also that the non-vanishing diagrams of figure 10.4 induce a logarithmically divergent correction to the kinetic energy term $\bar{\Phi}\Phi$. However, the diagrams of figure 10.5 vanish and no Φ^2 or $\bar{\Phi}^2$ term is produced by the radiative corrections. This may also be seen† by inverting the propagators corrected to 1-loop order to obtain the corrected two-point function. In the effective action we verify that no mass term shows up.

Figure 10.4

Figure 10.5

10.2.4 One-loop corrections to the vertices

We show that also no Φ^3 term is generated at the 1-loop level. For the massless case the corresponding diagram vanishes trivially while for the

† We may write the action as

$$\frac{1}{2!}\int \mathrm{d}^8z\,\mathrm{d}^8z'\,(S(z),\bar{S}(z))\,\Gamma^{(2)}(z,z')\begin{pmatrix}S(z')\\ \bar{S}(z')\end{pmatrix}$$

where $\Gamma^{(2)}(z,z') = A(z)\delta^8(z-z')$ and A is a 2×2 matrix of differential operators. Defining $A(z)\Delta(z,z') = \mathrm{i}\delta^8(z-z')$ we find $\int \mathrm{d}^8z''\Gamma^{(2)}(z',z'')\,\Delta(z'',z) = \mathrm{i}\delta^8\,(z-z')$. An appropriate regularisation must be used.

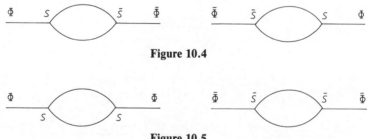

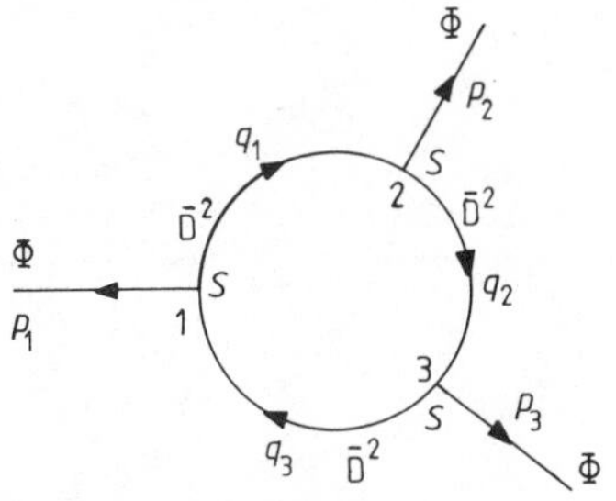

Figure 10.6

massive case (figure 10.6) we note, for example, that

$$\bar{D}_1^2 \Delta^{SS}(z_1, z_2) = \frac{\bar{m}}{4\Box(\Box - |m|^2)} \bar{D}_1^2 D_2^2 \delta^8(z_1 - z_2)$$

$$= \frac{\bar{m}}{4\Box(\Box - |m|^2)} 16 \int \frac{d^4 q_1}{(2\pi)^4}$$

$$\times \exp[-(\theta_1 \not{q}_1 \bar{\theta}_1 + \theta_2 \not{q}_1 \bar{\theta}_2 - 2\theta_1 \not{q}_1 \bar{\theta}_2)] \exp[i q_1(x_1 - x_2)] \tag{10.28}$$

and

$$\Phi(x, \theta, \bar{\theta}) = e^{i\theta\partial\bar{\theta}} \Psi(x, \theta) = \int \frac{d^4 p}{(2\pi)^4} e^{-\theta \not{p} \bar{\theta}} \tilde{\Psi}(p, \theta) e^{ipx}. \tag{10.29}$$

Hence the integrand contains all the $\bar{\theta}$ dependence in the exponential of the following expression

$$[-\theta_1 \not{p}_1 \bar{\theta}_1 - \theta_1 \not{q}_1 \bar{\theta}_1 - \theta_2 \not{q}_2 \bar{\theta}_2 + 2\theta_1 \not{q}_1 \bar{\theta}_2 - \theta_2 \not{p}_2 \bar{\theta}_2 - \theta_2 \not{q}_2 \bar{\theta}_2 - \theta_3 \not{q}_2 \bar{\theta}_3$$
$$+ 2\theta_2 \not{q}_2 \bar{\theta}_3 - \theta_3 \not{p}_3 \bar{\theta}_3 - \theta_3 \not{q}_3 \bar{\theta}_3 - \theta_1 \not{q}_3 \bar{\theta}_1 + 2\theta_3 \not{q}_3 \bar{\theta}_1]$$
$$= -\theta_1(\not{p}_1 + \not{q}_1 + \not{q}_3)\bar{\theta}_1 - \theta_2(\not{q}_1 + \not{q}_2 + \not{p}_2)\bar{\theta}_2 - \theta_3(\not{q}_2 + \not{p}_3 + \not{q}_3)\bar{\theta}_3$$
$$+ 2(\theta_1 \not{q}_1 \bar{\theta}_2 + \theta_2 \not{q}_2 \bar{\theta}_3 + \theta_3 \not{q}_3 \bar{\theta}_1)$$
$$= 2[(\theta_3 - \theta_1)\not{q}_3 \bar{\theta}_1 + (\theta_1 - \theta_2)\not{q}_1 \bar{\theta}_2 + (\theta_2 - \theta_3)\not{q}_2 \bar{\theta}_3] \tag{10.30}$$

where we use the four-momentum conservation relations $q_1 = p_2 + q_2$, $q_2 = q_3 + p_3$ and $q_3 = q_1 + p_1$. Since we have to integrate over the full superspace of θ_1, θ_2 and θ_3 the non-vanishing contribution comes from the coefficient of $(\theta_1^2 \theta_2^2 \theta_3^2 \bar{\theta}_1^2 \bar{\theta}_2^2 \bar{\theta}_3^2)$. In our case it must derive in view of (10.30) from the following product:

$$(\theta_3 - \theta_1)^2 (\theta_1 - \theta_2)^2 (\theta_2 - \theta_3)^2 \bar{\theta}_1^2 \bar{\theta}_2^2 \bar{\theta}_3^2 \, \tilde{\Psi}\tilde{\Psi}\tilde{\Psi} \tag{10.31}$$

which vanishes identically in view of the fact that $\delta^2(\theta - \theta') = (\theta - \theta')^2$ and $[\delta^2(\theta - \theta')]^2 = 0$. The same result could also be derived by the usual method of integration by parts and the vanishing result obviously holds for a $\bar{\Phi}^3$ cube vertex as well.

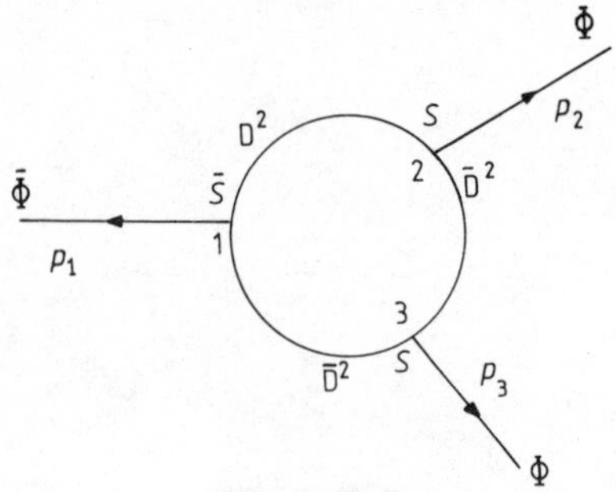

Figure 10.7

A non-vanishing $\bar{\Phi}\Phi^2$ vertex in the massive case is, however, induced by 1-loop corrections. With the self-explanatory notation shown in figure 10.7 the contribution apart from overall numerical factors is given by $(z_{12} = z_1 - z_2)$

$$\Gamma \sim \int d^8z_1 \, d^8z_2 \, d^8z_3 \, \bar{\Phi}(z_1)\Phi(z_2)\Phi(z_3)$$

$$\times \frac{D^2(z_1)}{(\Box_1 - m^2)} \delta^8(z_{12}) \frac{\bar{D}^2(z_2)D^2(z_2)}{\Box_2(\Box_2 - m^2)} \delta^8(z_{23}) \frac{\bar{D}^2(z_3)}{(\Box_3 - m^2)} \delta^8(z_{31}). \tag{10.32}$$

Integrating by parts on z_3, $(\bar{D}(z_3)\Phi_3 = 0)$, commuting $D^2, \bar{D}^2$ at different points in superspace and using $\bar{D}^2(z_3)\delta^8(z_{23}) = \bar{D}^2(z_2)\delta^8(z_{23})$, $\bar{D}^2D^2\bar{D}^2 = 16\Box\bar{D}^2$, we obtain

$$\Gamma \sim \int \ldots \left(\frac{D^2(z_1)}{(\Box_1 - m^2)} \delta^8(z_{12}) \frac{16\Box_2\bar{D}^2(z_2)}{\Box_2(\Box_2 - m^2)} \delta^8(z_{23}) \frac{1}{(\Box_3 - m^2)} \right.$$

$$\left. \times \delta^4(x_{31}) \, \delta^4(\theta_{31}) \right). \tag{10.33}$$

Another integration by parts now on z_2 gives $(\bar{D}(z_2)\Phi(z_2) = 0)$

$$\Gamma \sim \int \ldots \left(\frac{\bar{D}^2(z_2)D^2(z_1)}{(\Box_1 - m^2)} \delta^8(z_{12})\delta^4(\theta_{23})\delta^4(\theta_{31}) \right.$$

$$\left. \times \frac{16\Box_2}{\Box_2(\Box_2 - m^2)} \delta^4(x_{23}) \frac{1}{(\Box_3 - m^2)} \delta^4(x_{31}) \right) \tag{10.34}$$

$$\sim \int \ldots \left[\frac{16}{(\Box_1 - m^2)} \delta^4(x_{12}) \right] \delta^4(\theta_{12})\delta^4(\theta_{31}) \frac{16\Box_2}{\Box_2(\Box_2 - m^2)}$$

$$\times \delta^4(x_{23}) \frac{1}{(\Box_3 - m^2)} \delta^4(x_{31}). \tag{10.35}$$

The result after integration over θ_2, θ_3 is local in $\theta, \bar{\theta}$ space

$$\Gamma \sim \int d^4\theta_1 \, d^4x_1 \, d^4x_2 \, d^4x_3 \, \bar{\Phi}(1)\Phi(2)\Phi(3)$$

$$\times \left(\frac{16}{\Box_1 - m^2} \delta^4(x_{12}) \right) \left(\frac{16\Box_2}{\Box_2(\Box_2 - m^2)} \delta^4(x_{23}) \right) \left(\frac{1}{\Box_3 - m^2} \delta^4(x_{31}) \right) \tag{10.36}$$

which clearly does not vanish on integration over θ_1. We may go to the momentum space by introducing the Fourier transforms of the Φ's and delta functions. The degree of divergence of the diagram is (-2) rather than its superficial value (-4) due to the $\Box_2$ in the numerator which was introduced during the manipulation of $\mathrm{D}, \bar{\mathrm{D}}$ operators. The explicit calculation here reinforces the validity of the manipulations in the momentum space of $\mathrm{D}, \bar{\mathrm{D}}$ operators, as done earlier.

To illustrate a *2-loop calculation* and the power of superfield graphs we consider figure 10.8 which may generate a correction to the Φ^3 term in the effective potential.

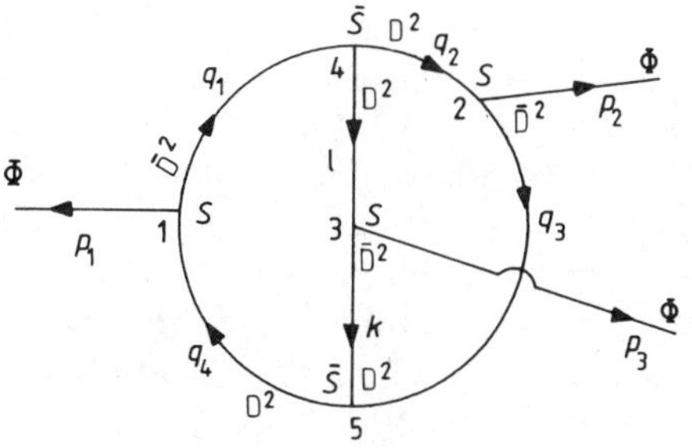

Figure 10.8

Making use of (10.29) and

$$\begin{aligned} \mathrm{D}_1^2\delta^4(\theta_{12}) &= -4\delta^2(\bar{\theta}_{12})\exp(\theta_1-\theta_2)\not{k}\bar{\theta}_1 \\ \bar{\mathrm{D}}_1^2\delta^4(\theta_{12}) &= -4\delta^2(\theta_{12})\exp\theta_1\not{k}(\bar{\theta}_2-\bar{\theta}_1) \end{aligned} \tag{10.37}$$

derived from (8.27), when they act on $\exp[\mathrm{i}k(x_1-x_2)]$ we find that all the $\bar{\theta}_i$ dependence of the diagram is contained in the following expression:

$$\begin{aligned} &\delta^2(\theta_{14})\delta^2(\theta_{25})\delta^2(\bar{\theta}_{42})\delta^2(\bar{\theta}_{51})\delta^2(\bar{\theta}_{43}) \\ &\times\exp[(\theta_4-\theta_2)\not{q}_2\bar{\theta}_4+(\theta_4-\theta_3)\not{l}\bar{\theta}_4+(\theta_5-\theta_1)\not{q}_4\bar{\theta}_5 \\ &-\theta_1\not{q}_1(\bar{\theta}_1-\bar{\theta}_4)-\theta_2\not{q}_3(\bar{\theta}_2-\bar{\theta}_5)-(\theta_5-\theta_3)\not{k}\bar{\theta}_5 \\ &-\theta_3\not{k}(\bar{\theta}_3-\bar{\theta}_5)-\theta_1\not{p}_1\bar{\theta}_1-\theta_2\not{p}_2\bar{\theta}_2-\theta_3\not{p}_3\bar{\theta}_3]. \end{aligned} \tag{10.38}$$

The argument of the exponential is simplified owing to the presence of the delta functions over θ's and conservation of four-momentum at each vertex. We find for $p_1=p_2=p_3=0$

$$[\ldots]=[2(\theta_1-\theta_2)\not{q}_3-2(\theta_3-\theta_1)\not{k}](\bar{\theta}_2-\bar{\theta}_1) \tag{10.39}$$

All the $\bar{\theta}$ dependence is removed after integration on $\bar{\theta}_2, \bar{\theta}_3, \bar{\theta}_4, \bar{\theta}_5$. The integration over $\bar{\theta}_1$ gives a vanishing result. It would be at best tedious to perform the calculation in component field formulation.

10.2.5 Non-renormalisation theorem

With the unconstrained superpotentials introduced in §9.3 the superspace perturbation theory requires that all the vertices (even in the presence of gauge superfields) are to be integrated over the full superspace, that is, over $\int d^4x \int d^4\theta$. Moreover, as discussed in Chapter 8 we may then freely integrate by parts to move within any loop the covariant derivatives on any propagator off the corresponding $\delta^4(\theta_k - \theta_l)$ until only one fermionic integration is left. Proceeding analogously for all the loops in a diagram the final result is an expression of the type

$$\int d^4x_1 \dots d^4x_n \, d^4\theta F_1(x_1, \theta, \bar{\theta}) \dots F_n(x_n, \theta, \bar{\theta}) g(x_1, \dots, x_n) \qquad (10.40)$$

where $F_i(x_i, \theta, \bar{\theta})$ are products of superfields (and their derivatives) and g is translationally invariant. The final result is thus invariant under translations in superspace. Counter terms that cannot be rewritten locally as the full supersymmetric integral (10.40) cannot arise in perturbation theory. In the Wess–Zumino model, for example, the mass and interaction terms cannot occur as counter terms†. In other words the superpotential of chiral superfields is not renormalised. The result (10.40) is the statement of the non-renormalisation theorem of $N = 1$ supersymmetry.

A counter term of the form $Z_{ij}\bar{\Phi}_i\Phi_j$, the kinetic term, is allowed and leads to wave function renormalisation $(\sqrt{\mathbf{Z}})_{ij}\Phi_j = \Phi_i^{(0)}$ where '0' indicates the bare quantities and $\sqrt{\mathbf{Z}}$ indicates the positive square root of the Hermitian positive matrix $\mathbf{Z} = (Z_{ij})$. From $m_{ij}^{(0)} \Phi_i^{(0)}\Phi_j^{(0)} = m_{ij}\Phi_i\Phi_j$, $g_{ijk}^{(0)}\Phi_i^{(0)}\Phi_j^{(0)}\Phi_k^{(0)} = g_{ijk}\Phi_i\Phi_j\Phi_k$ we obtain $m_{ij} = m_{i'j'}^{(0)}(\sqrt{\mathbf{Z}})_{i'i}(\sqrt{\mathbf{Z}})_{j'j}$ and $g_{ijk} = g_{i'j'k'}^{(0)}(\sqrt{\mathbf{Z}})_{i'i}(\sqrt{\mathbf{Z}})_{j'j}(\sqrt{\mathbf{Z}})_{k'k}$. The supersymmetry of the action thus imposes constraints relating the wave function, the coupling constant and the multiplicative mass renormalisation constants. In the case when only chiral superfields are present we need to perform only a wave function renormalisation. In conventional theories the coupling constant is renormalised independently of the wave function and any renormalisation prescription for the first one cannot influence the kinetic energy terms which are controlled by the second. In SS theories it is then clear that the kinetic term dictates the correct renormalisation prescription so as to avoid negative kinetic energy (Amati and Chou 1982) and the renormalisation of the coupling constant then follows.

We will now illustrate the non-renormalisation theorem for the Wess–Zumino model in the background field method. We split Φ into a background and a quantum part (indicated for convenience by Φ itself):

$$\Phi \to \Phi_b + \Phi \qquad \bar{D}\Phi = \bar{D}\Phi_b = 0. \qquad (10.41)$$

† Barring that terms like $\int d^4\theta \, \Phi^2(D^2/\Box) \, \Phi$ are not produced in the effective action (Grisaru *et al.* 1979).

In the background field method we compute the one-particle-irreducible (OPI) graphs with only external background legs to calculate in the perturbation theory the background effective action $\Gamma[\Phi_b]$.

On substituting $(\Phi+\Phi_b)$ in place of Φ the background action for the Wess–Zumino model is

$$I_b = \int d^8z\bar{\Phi}_b\Phi_b + \left(\int d^6s(\tfrac{1}{2}m\Phi_b^2 + \tfrac{1}{3}g\Phi_b^3) + \text{cc}\right) \qquad (10.42)$$

which is just the tree-level contribution to $\Gamma[\Phi_b]$. From the terms quadratic in the quantum part Φ

$$\int d^8z\bar{\Phi}\Phi + \left(\int d^6s(m+2g\Phi_b)\Phi^2 + \text{cc}\right) \qquad (10.43)$$

we derive the propagators and the background-quantum vertices with two quantum legs. The terms of higher order in Φ, i.e. the Φ^3 term in our case, occur as vertices in two- and higher-loop diagrams. Terms linear in Φ or involving only the background field are not used in the Feynman rules for calculating the quantum corrections to the effective action.

Introducing the unconstrained superfields $\Phi = -\frac{1}{4}\bar{D}^2S$, $\bar{\Phi} = -\frac{1}{4}D^2\bar{S}$ discussed in §9.3 the Feynman rules for the background-quantum vertices are derived from the following term:

$$g\int d^8zS(-\tfrac{1}{4}\bar{D}^2S)\Phi_B + \text{cc}. \qquad (10.44)$$

Adding to the action the gauge-fixing term (9.18) we derive the free propagators given in (9.27). The perturbation theory calculation may now be done using (9.15) and (10.44). We note that the background field Φ_B is not required to be expressed in terms of the unconstrained superfield to obtain (10.44) as an integral over the full superspace. The non-renormalisation theorem is now apparent. The Feynman rules derived above give contributions to the effective action that involve only full superspace integrals and couplings to the constrained background superfield Φ_b. If we admit only local operators it is not possible to write the mass and interaction counter terms depending only on Φ_b as full superspace integrals and consequently they are ruled out.

The above result has been confirmed up to two loops by an explicit computation of the effective potential using the superfield tadpole or bubble methods (see Srivastava 1984, 1985).

We mention in passing that a suitable regularisation procedure which preserves supersymmetry must be used in the quantised theory. In practice dimensional reduction regularisation, which seems to guarantee SS in the counter terms at least at low-loop orders, is used (Siegel 1979, 1980). However, care must be taken in massless theories to separate the ultraviolet from the infrared divergences which themselves require some form of infrared regularisation. For the Wess–Zumino model we may use Pauli–Villars–Gupta or higher covariant derivative regularisations (Illiopoulos and Zumino 1974, Ferrara and Piguet 1975).

The derivation of a non-renormalisation theorem for the simple and N-extended SS gauge theories involves additional complications due to the presence of ghost terms and the problem of 'ghosts for ghosts'. We refer the reader to the current literature (Stelle 1984 and references cited therein).

10.3 Abelian gauge theory. Examples

We now give some examples from the theory of massless chiral superfields interacting with an Abelian gauge superfield with the action given in (5.28). The interaction term apart from the self-interaction of Φ is given by ($g_1 = 1$)

$$H_{\text{int}} = -\int \mathrm{d}^8 z\, \bar{\Phi}(\mathrm{e}^{QV} - 1)\Phi = -\int \mathrm{d}^8 z (-\tfrac{1}{4}\mathrm{D}^2\bar{S})\left(QV + \frac{1}{2!}Q^2V^2 + \frac{1}{3!}Q^3V^3 + \ldots\right)(-\tfrac{1}{4}\bar{\mathrm{D}}^2 S) \tag{10.45}$$

and the Feynman rules may be read from it.

10.3.1 *One-loop self-energy of gauge superfield*

We consider first the graph of figure 10.9.

$$\Gamma \sim \frac{(-\mathrm{i})^2}{2!}(-Q)^2 \int \mathrm{d}^8 z_1\, \mathrm{d}^8 z_2 V(z_1)V(z_2)\left(\frac{\mathrm{i}}{16\Box_1}\mathrm{D}^2(z_1)\delta^8(z_{12})\overleftarrow{\bar{\mathrm{D}}}{}^2(z_2)\right) \times \left(\frac{\mathrm{i}}{16\Box_2}\mathrm{D}^2(z_2)\delta^8(z_{21})\overleftarrow{\bar{\mathrm{D}}}{}^2(z_1)\right)$$

$$= \frac{Q^2}{2}\int \mathrm{d}^8 z_1\, \mathrm{d}^8 z_2 V(1)V(2)\left(\frac{1}{\Box_1}\frac{\mathrm{D}_1^2\bar{\mathrm{D}}_1^2}{16}\delta^8(z_{12})\right)\left(\frac{1}{\Box_2}\frac{\bar{\mathrm{D}}_1^2\mathrm{D}_1^2}{16}\delta^8(z_{12})\right). \tag{10.46}$$

Integration by parts on z_1 gives ($\delta \equiv \delta^8(z_{12}), \delta^4 = \delta^4(x_{12})$):

$$\begin{aligned}
[\mathrm{D}^2\bar{\mathrm{D}}^2\delta]\,[\bar{\mathrm{D}}^2\mathrm{D}^2\delta]\,V &\to [\bar{\mathrm{D}}^2\delta]\,\mathrm{D}^2[(\bar{\mathrm{D}}^2\mathrm{D}^2\delta)V] \\
&\to \delta\bar{\mathrm{D}}^2[(\mathrm{D}^2\bar{\mathrm{D}}^2\mathrm{D}^2\delta) + 2(\mathrm{D}^\alpha\bar{\mathrm{D}}^2\mathrm{D}^2\delta)\mathrm{D}_\alpha + (\bar{\mathrm{D}}^2\mathrm{D}^2\delta)\mathrm{D}^2]\,V \\
&= \delta\bar{\mathrm{D}}^2[(16\Box\,\mathrm{D}^2\delta) - 8\mathrm{i}(\partial^\alpha{}_{\dot{\alpha}}\bar{\mathrm{D}}^{\dot{\alpha}}\mathrm{D}^2\delta)\mathrm{D}_\alpha + (\bar{\mathrm{D}}^2\mathrm{D}^2\delta)\mathrm{D}^2]\,V \\
&= \delta[(16\Box\,\bar{\mathrm{D}}^2\mathrm{D}^2\delta) + 16\mathrm{i}(\partial^\alpha{}_{\dot{\alpha}}\bar{\mathrm{D}}_{\dot{\beta}}\bar{\mathrm{D}}^{\dot{\alpha}}\mathrm{D}^2\delta)\bar{\mathrm{D}}^{\dot{\beta}}\mathrm{D}_\alpha + (\bar{\mathrm{D}}^2\mathrm{D}^2\delta)\bar{\mathrm{D}}^2\mathrm{D}^2]\,V \\
&= \delta[(16^2\Box\delta^4) + 16^2\mathrm{i}\cdot\tfrac{1}{2}\cdot(\partial^\alpha{}_{\dot{\beta}}\delta^4)\bar{\mathrm{D}}^{\dot{\beta}}\mathrm{D}_\alpha + (16\delta^4)\bar{\mathrm{D}}^2\mathrm{D}^2]\,V
\end{aligned} \tag{10.47}$$

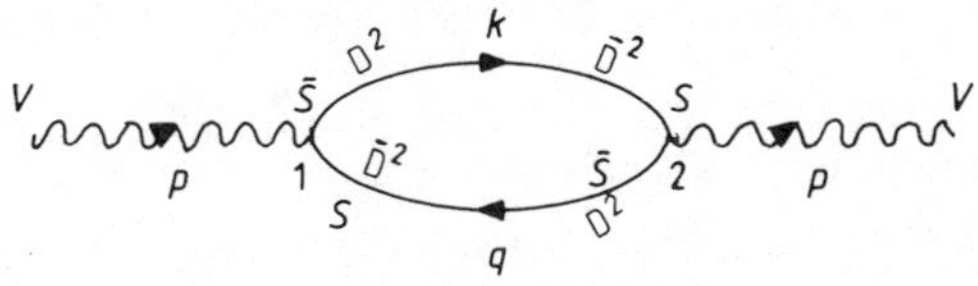

Figure 10.9

where we have made use of (3.23), (3.25) and (8.39), (8.40) to drop several terms on sight.

Hence we obtain in momentum space ($q = k - p$):

$$\Gamma \sim \frac{Q^2}{2}\int \frac{d^4p}{(2\pi)^4}\, d^4\theta_1 \tilde{V}(p,\theta_1,\bar{\theta}_1)$$

$$\times \left[\int \frac{d^4k}{(2\pi)^4} \frac{[-q^2 + \frac{1}{2}q^{\alpha}{}_{\dot{\beta}}\bar{D}^{\dot{\beta}}D_{\alpha} + \frac{1}{16}\bar{D}^2D^2]\,\tilde{V}(-p,\theta_1,\bar{\theta}_1)]}{k^2(k-p)^2}\right] \qquad (10.48)$$

where $D_{\alpha} = D_{\alpha}(-p,\theta_1,\bar{\theta}_1)$, $\bar{D}_{\dot{\alpha}} = (-p,\theta_1,\bar{\theta}_1)$ do not depend on k.

We consider next the contribution from the tadpole graph of figure 10.10. We find

$$\Gamma \sim (-i)\frac{(-Q^2)}{2!}\int d^8z \left[V(z)V(z')\frac{iD^2\bar{D}^2}{16\Box}\delta^8(z-z')\right]_{z=z'}$$

$$= -\frac{Q^2}{2}\int d^8z \left[\int \frac{d^4p\, d^4p'}{[(2\pi)^4]^2}\tilde{V}(-p,\theta,\bar{\theta})\tilde{V}(p',\theta,\bar{\theta})\, e^{i(p'-p)x}\right.$$

$$\left.\times \int \frac{d^4k}{(2\pi)^4}\left(\frac{-1}{k^2}\right)\right]$$

$$= -\frac{Q^2}{2}\int \frac{d^4p}{(2\pi)^4}\, d^4\theta \tilde{V}(p,\theta,\bar{\theta})\tilde{V}(-p,\theta,\bar{\theta}) \int \frac{d^4k}{(2\pi)^4}\left(\frac{-1}{k^2}\right). \qquad (10.49)$$

The quadratic divergence thus cancels and the final result is logarithmically divergent. The quadratically divergent integral in fact is zero when we use dimensional regularisation (see, for example, 't Hooft and Veltman(1973)). It is clear from this example that we may set up rules to do the algebraic manipulations of integration by parts on the diagrams themselves. They are very useful for higher-loop calculations.

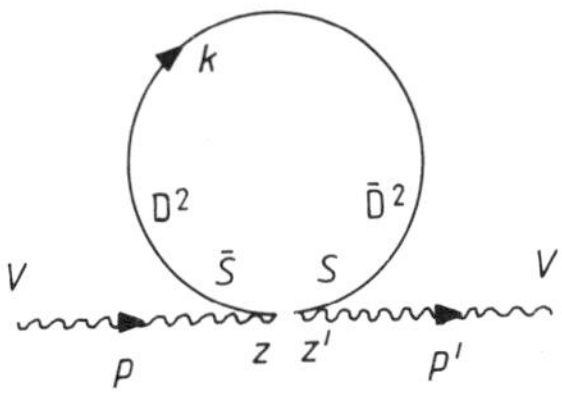

Figure 10.10

10.3.2 Radiative generation of D-term

We discussed in Chapter 6 the relevance of a D-term in inducing a spontaneous SS breaking. We investigate now the radiative generation of such a counter term by the V-tadpole supergraph (figure 10.11). The contribution at 1-loop order with chiral superfields inside is easily found:

$$\Gamma \sim \sum_i Q_i \int \mathrm{d}^8 z \left(\frac{\mathrm{D}_i^2(z)\bar{\mathrm{D}}_i^2(z')}{\Box - |m_i|^2} \delta^8(z-z') \right)_{z=z'} V(z)$$

$$\sim \left(\sum_i Q_i \right) \int \mathrm{d}^4\theta\, \tilde{V}(0,\theta,\bar{\theta}) \int \frac{\mathrm{d}^4 k}{(2\pi)^4} \frac{1}{(k^2 + |m_i|^2)} \tag{10.50}$$

where

$$\mathrm{Tr}\, \mathbf{Q} = \sum_i Q_i$$

is the trace or sum of all the U(1) charges of the chiral fields involved. We obtain a quadratically divergent integral and the contribution is proportional to (Tr $\mathbf{Q}$). It is sufficient therefore that Tr $\mathbf{Q} = 0$ for a vanishing D-term. This is true, for example, in SSQED where the gauge invariance requires that there be two chiral superfields of opposite charges; this is usually true in grand unified theories. For massless chiral fields the quadratic integral itself vanishes when using dimensional regularisation. For the massive case the gauge-invariant Pauli–Villars regulator chiral superfields do not contribute either. It has been shown recently (Fischler *et al.* (1983); see also Ovrut and Wess (1982)) that the higher order perturbation theory also does not generate a Fayet–Illiopoulous term if Tr $\mathbf{Q} = 0$.

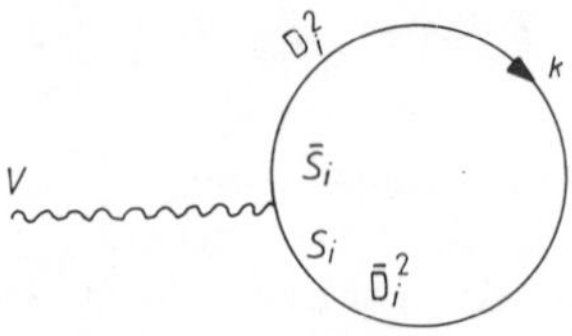

Figure 10.11

10.4 Effective potential

Superfields are also a very economical tool for computing the effective potential in an SS theory. We shall discuss here for illustration the superfield tadpole method of calculating the effective potential in the Wess–Zumino model.

We shift the chiral superfield by a constant (background) chiral superfield $C(\theta) = a + f\theta^2$ with vanishing spinor (or tensor) component. The free action of the shifted theory is now given by

$$\int d^8z\bar{\Phi}\Phi + \left(\left(\int d^6sM(\theta)\Phi^2 + \text{cc}\right)\right) \tag{10.51}$$

where the 'mass' $M(\theta) = (m + 2ga) + 2g\,f\theta^2 = \tilde{a} + \tilde{f}\theta^2$ is now a superfield while the interaction term is found to be

$$\int d^6s[\tfrac{1}{3}\,g\Phi^3 + mC(\theta)\Phi + gC^2(\theta)\Phi] + \int d^8z\bar{C}(\bar{\theta})\Phi + \text{cc}. \tag{10.52}$$

We may add the gauge-fixing term (9.18) to (10.51) and derive the propagators of the shifted theory. They are found to be ((9.44), (9.47)):

$$\Delta^{s\bar{s}} = \left(\frac{1}{\xi}\frac{P_2 + P_{\mathrm{T}}}{\Box} + P_1(\Box - \bar{M}P_1M)^{-1}\right)\delta^8(z - z')$$

$$\Delta^{\bar{s}\bar{s}} = \frac{M(\theta)}{4\Box}\bar{\mathrm{D}}^2\Delta^{s\bar{s}} \tag{10.53}$$

and fortunately the term $P_1(\Box - \bar{M}P_1M)^{-1}$ can be expanded (even for n interacting chiral superfields) in a compact form exhibiting the pole structure, while the θ, $\bar{\theta}$ dependence is separated out. In our case (suppressing the ξ dependent term)

$$\Delta^{s\bar{s}} = \frac{P_1}{\Box - |\tilde{a}|^2}\delta^8(z - z')$$
$$+ \exp(-\mathrm{i}\theta\partial\!\!\!/\bar{\theta} - \mathrm{i}\theta'\partial\!\!\!/\bar{\theta}')(A\bar{\theta}^2\theta'^2 + B\Box^{-2} + C\Box^{-1}\bar{\theta}^2 + E\Box^{-1}\theta'^2]\delta^4(x - x') \tag{10.54}$$

where

$$A = \frac{|\tilde{f}|^2}{(\Box - |\tilde{a}|^2)[(\Box - |\tilde{a}|^2)^2 - |\tilde{f}|^2]} \qquad B = |\tilde{a}|^2A$$

$$C = E^* = \frac{\tilde{a}\tilde{f}^*}{(\Box - |\tilde{a}|^2)^2 - |\tilde{f}|^2}. \tag{10.55}$$

The superfield tadpole method (Srivastava 1983*b*) consists simply in computing the one-point functions of the shifted theory to the desired number of loops. It gives directly the first partial derivatives of the effective potential which are, for example, needed to discuss spontaneous symmetry breaking or imposing renormalisation constraints.

In the zero-loop approximation we read from (10.52)

$$\Gamma_0^{(1)} = \int d^2\theta\left(mC(\theta) + gC^2(\theta) + \int d^2\bar{\theta}\bar{C}(\bar{\theta})\right)\tilde{\Phi}(0, \theta, \bar{\theta}) + \text{cc} \tag{10.56}$$

where $\tilde{\Phi}(p, \theta, \bar{\theta})$ is the Fourier transform of Φ. We find

$$\Gamma_0^{(1)} = \tilde{a}f\tilde{A}(0) + (ma + ga^2 + f^*)\tilde{F}(0) + \text{cc} \tag{10.57}$$

and we may read off the tadpole contributions of the component fields to obtain

$$-\frac{\partial V_0}{\partial a} = \tilde{a}f \qquad -\frac{\partial V_0}{\partial f} = f^* + a(m + ga). \tag{10.58}$$

On integration we obtain the tree-level result

$$-V_0 = \left(f^* f + f\frac{\partial W}{\partial a} + \bar{f}\frac{\partial \overline{W}}{\partial \bar{a}}\right) \tag{10.59}$$

where $W = \frac{1}{2}ma^2 + \frac{1}{3}ga^3$. The (scalar) effective potential is obtained by $a \to A_c$, $f \to F_c$ where A_c and F_c are classical fields. We find $V_0 = |F_c|^2$ if we use auxiliary field equations of motion.

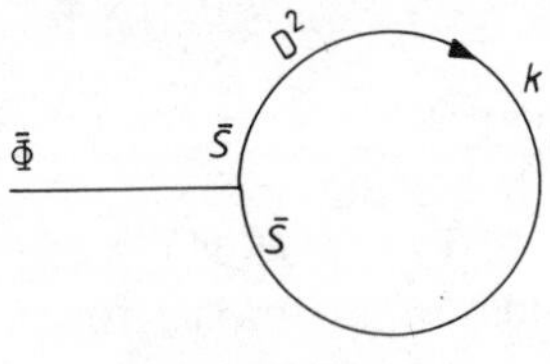

Figure 10.12

To obtain the 1-loop contribution we need to compute, unlike in the case of component formulation, just one superfield tadpole graph (figure 10.12). We find

$$\begin{aligned}
\mathrm{i}\Gamma_1^{(1)} &= \mathrm{i}g \int \mathrm{d}^8 z \bar{\Phi}(z)[-\tfrac{1}{4}\mathrm{D}^2\Delta^{\bar{S}\bar{S}}(z,z')]_{z=z'} \\
&= \mathrm{i}g \int \mathrm{d}^4\theta\, \tilde{\bar{\Phi}}(0,\theta,\bar{\theta})[-\tfrac{1}{4}\mathrm{D}^2\Delta^{\bar{S}\bar{S}}(z,z')]_{z=z'} \\
&= \int \frac{\mathrm{d}^4 k}{(2\pi)^4}\left(g\tilde{f}\tilde{\bar{F}}(0) - \frac{2\tilde{a}g|\tilde{f}|^2}{k^2 + |\tilde{a}|^2}\tilde{\bar{A}}(0)\right)\frac{1}{[(k^2 + |\tilde{a}|^2)^2 - |\tilde{f}|^2]}
\end{aligned} \tag{10.60}$$

and consequently

$$\begin{aligned}
-\frac{\partial V_1}{\partial f^*} &= -\mathrm{i}\int \frac{\mathrm{d}^4 k}{(2\pi)^4}\frac{g\tilde{f}}{[(k^2 + |\tilde{a}|^2)^2 - |\tilde{f}|^2]} \\
-\frac{\partial V_1}{\partial a^*} &= \mathrm{i}\int \frac{\mathrm{d}^4 k}{(2\pi)^4}\frac{2g\tilde{a}|\tilde{f}|^2}{[(k^2 + |\tilde{a}|^2)^2 - |\tilde{f}|^2](k^2 + |\tilde{a}|^2)}
\end{aligned} \tag{10.61}$$

plus their complex conjugates. Integrating we obtain

$$V_1 = \frac{-\mathrm{i}}{2}\int \frac{\mathrm{d}^4 k}{(2\pi)^4}\ln\left(1 - \frac{|\tilde{f}|^2}{k^2 + |\tilde{a}|^2}\right). \tag{10.62}$$

For $f = 0$ it vanishes showing the general result in an SS theory: if the SS is not broken at the tree level the radiative corrections will not break it either. The superfield bubble and the background superfield methods are among other alternative ways of calculating the effective potential.

Problems

10.1 Compute the corrections to the SS propagator in the chain approximation.

10.2 Compute from (10.22)(or(10.26)) the multiplicative mass and wave function renormalisation constants to 1-loop order ($\Phi = \sqrt{Z}\Phi_r$, $m_r = Z_m m$). Use an appropriate regularisation for the divergent integrals and find a relation between them.

11

SUPERGRAVITY

11.1 Introduction

Rigorous rigid supersymmetry, i.e. exact supersymmetry of the Lagrangian and the vacuum, implies degeneracy among bosonic and fermionic energy levels. Applying an SS generator to any bosonic (fermionic) state which is not annihilated by it we create a fermionic (bosonic) state with the same energy and momentum. All the known elementary particles should consequently have superpartners with the same mass. No such mass degeneracy is observed in nature. We list in table 11.1 some of the superpartners required when we attempt to build a supersymmetric unified theory of fundamental interactions (see, for example, Ellis 1985). If supersymmetry is to be relevant for the physical world it must be broken spontaneously or softly so that some of the superpartners may be made massive to be in agreement with the present experiments (Ellis 1984, 1985), for example, $m_{\tilde{q}}$, $m_{\tilde{l}}$, $m_{\tilde{W}}$, $m_{\tilde{H}} \sim$ O(15–20) GeV.

Table 11.1 Superpartners of some particles required for a super-unified theory of fundamental interactions.

Particle	Spin	Sparticle	Spin
Quark q_L, q_R	1/2,1/2	Sqark $\tilde{q}_L, \tilde{q}_R$	0,0
Lepton l_L, l_R	1/2,1/2	Slepton $\tilde{l}_L, \tilde{l}_R$	0,0
Photon γ	1	Photino $\tilde{\gamma}$	1/2
Gluon g	1	Gluino $\tilde{g}$	1/2
$W^{+,-}$	1	W-ino $\tilde{W}^{+,-}$	1/2
Z^0	1	Z-ino $\tilde{Z}^0$	1/2
Higgs H	0	Shiggs $\tilde{H}$	1/2

We found, however, in Chapter 6 that when global supersymmetry is realised in the spontaneously broken mode in which the SS generators do not annihilate the vacuum there appears in the theory a massless spin-1/2 Majorana fermion ('goldstino') field. It is not possible to identify, for example, the 'goldstino' field with the electron neutrino because it would satisfy low energy theorems which contradict the observed properties of the neutrino spectrum. The problem of the apparent non-existence of the 'goldstino' particle may be overcome by lifting the global supersymmetry to a local one where the transformation parameters become space–time dependent and a spin-3/2 compensating gauge field ψ_μ has to be added in the theory. The goldstino field in this context assumes the role of a gauge degree of freedom and a gauge can be chosen which eliminates it entirely from the theory while the gauge field ψ_μ itself become massive—the super-Higgs effect—analogous to the well known standard Higgs effect of ordinary gauge theories (see, for example, Deser and Zumino (1977), and references contained therein).

Among several other strong motivations for studying the local supersymmetry we may mention the problem of the observed smallness of the cosmological constant Λ which implies in our universe a very small value for the vacuum energy density $V_0 \simeq \Lambda \leqslant 10^{-120} M^4 \simeq (3\times 10^{-12}\ \text{GeV})^4$ where $M^{-1}=(8\pi G_N)^{1/2}$. In the case of global supersymmetry the vacuum energy is the order parameter for broken supersymmetry and it is positive definite. The vacuum energy density gets only positive contributions proportional to the supersymmetry breaking scale excluding the possibility of cancellations to obtain a small value if other constraints of particle phenomenology are also to be taken care of. On the other hand local supersymmetry (also called supergravity theory) necessarily brings gravitation into the theory and the vacuum energy has the possibility of becoming positive, negative or vanishing. This offers a hope for adjusting at the classical level a very small value for the cosmological constant which may not be spoiled by quantum corrections because of the residual symmetries in the theory. We will mention in §11.3 such a model which seems to follow also from the recent investigations on superstring models in higher space–time dimensions. Supersymmetry also offers some hope for resolving certain other unsolved problems in the field of cosmology as well.

11.2 Pure $N=1$ supergravity

11.2.1 Local supersymmetry

When we promote a rigid (global) symmetry with constant parameters ε to a local symmetry by letting† $\varepsilon \to \varepsilon(x)$ the kinetic terms of the action are no longer invariant and we find

$$\delta I = \int \mathrm{d}^4 x\, j^m_N \partial_m \varepsilon \tag{11.1}$$

where $j^m_N(x)$ is an on-shell conserved Noether's current of the global symmetry. In order to restore the symmetry we are required to introduce in the theory an additional compensating gauge field. We may cancel the

† See problem 2.6.

variation (11.1) by adding, apart from a kinetic term for the gauge field, a new interaction term (minimal coupling) to the action

$$I' = -\varkappa \int d^4x j_N^m A_m \qquad \delta A_m = \frac{1}{\varkappa}\,\partial_m \varepsilon + \ldots \tag{11.2}$$

where $\varkappa$ is a constant and the transformation law of the gauge field contains the derivative of the symmetry parameter. But a new term will arise from δj_N^m which again has to be compensated by adding more terms to the action and possibly to the earlier transformation rules under the local symmetry of the fields involved as well. Working step by step, the so called Noether procedure (which becomes an iterative procedure in $\varkappa$) may result in a locally symmetric action after a finite number of steps.

The procedure may be followed for supersymmetric theories as well. However, their transformation laws at the global level already contain the derivatives of the component fields. We assume that at the local level only the gauge fields transform with terms that are proportional to $\partial_m\varepsilon$ while the other fields contain $\varepsilon(x)$ but not $\partial_m\varepsilon(x)$, e.g $\delta\psi = -\sqrt{2}\,[\mathrm{i}(\sigma_m\bar{\varepsilon}(x))\partial^m A + F(x)\varepsilon]$. The spinor parameter carries dimension $-1/2$ and it follows that if we require dimension 3/2 for the vector–spinor fermionic gauge field indicated by ψ_m the parameter $\varkappa$ will carry dimension $[\varkappa] = -1$, $\delta\psi_m(x) = 1/\varkappa - \partial_m\varepsilon(x) + \ldots$. The necessity of the constant $\varkappa$ with non-vanishing dimension is a hint that gravity should enter in a locally supersymmetric theory. The gauge field of the local supersymmetry is a (real) Majorana field since $\delta\psi_m \sim \partial_m\varepsilon$ and ε is a Majorana spin-1/2 parameter. Since the Noether supercurrent carries dimension 7/2 we obtain the Noether coupling term $\varkappa(\bar{\psi}_m J_N^m)$. Having introduced new fermionic degrees of freedom we must also introduce more bosonic degrees of freedom to balance them so as to maintain Fermi–Bose supersymmetry. This also follows from the fact that the supersymmetry current transforms into the stress tensor T^{lm} of the matter system. The coupling $\bar{\psi}_m J_N^m$ thus requires at the same time a term $\varkappa h_{lm} T^{lm}$ where $h_{lm} = h_{ml}$ with dimension 1 is a new bosonic (spin-2) compensating gauge field (of linearised gravity). The gauge fields of local supersymmetry thus seem to belong to the (2, 3/2) supergravity supermultiplet of $N=1$ supersymmetry mentioned in Chapter 2 rather than the (3/2, 1) supermultiplet. We remark that the gravity (spin-2) is necessarily coupled to the stress tensor of all matter while neither the real spin-3/2 field nor any other real matter field can couple minimally to a spin-1 photon. The gauge field ψ_m, being a superpartner of the spin-2 graviton, should describe a spin-3/2 particle, which is called the gravitino. The simplest locally supersymmetric theory contains just these two fields and will be described below after a brief sketch of the tetrad formulation of ordinary gravity, which is needed since we have fermions in the theory.

11.2.2 Tetrad formulation of Einstein–Cartan theory of gravitation†

The space–time manifold is labelled by coordinates x^μ where $\mu = 0, 1, 2, 3$ indicates the world index. We may introduce at each space–time point a sufficiently differentiable field of four vectors (vierbein or tetrad frame) $e_m = e_m{}^\mu \partial_\mu$ where $m = 0, 1, 2, 3$ and $[e_\mu^m] = 0$. We assume also the existence of a constant Minkowski metric η_{mn} and choose the tetrads to be orthonormal, $e_m \cdot e_n = \eta_{mn}$. We have the dual frame $e^m = e_\mu^m \, dx^\mu$ and find $e_\mu^n e_m^\mu = \delta_\nu^\mu e_\mu^n e_m^\nu = \delta_m^n$ which implies $e_\mu^m e_m^\nu = \delta_\mu^\nu$. We may define anholonomic components of a tensor field referred to a tetrad basis, for example, $A^m = e_\mu^m A^\mu$, $\eta_{nm} = e_n^\mu e_m^\nu g_{\mu\nu}$, $A^m B_m = e_\nu^m A^\mu e_m^\nu B_\nu = A^\mu B_\mu$ etc. The general coordinate transformations $x_\mu \to x'_\mu$ or diffeomorphisms keep the local tetrad frames fixed while the local Lorentz transformations describe the rotations of the tetrad frames, independently from each other at each point x. The tetrad fields $e_\mu^m(x)$ are supposed to describe gravitation. Under a combined infinitesimal transformation we find

$$\begin{aligned} \delta e_\mu^m &= \zeta^\alpha(x)\partial_\alpha e_\mu^m + (\partial_\mu \zeta^\nu) e_\nu^m + \lambda^m{}_n e_\mu^n \\ \delta e_m^\mu &= \zeta^\alpha(x)\partial_\alpha e_m^\mu - (\partial_\nu \zeta^\mu) e_m^\nu - \lambda^n{}_m e_n^\mu \end{aligned} \tag{11.3}$$

where $x'_\mu \simeq x_\mu - \zeta_\mu(x)$ and the Lorentz rotation parameters satisfy $\lambda^m{}_n = -\lambda_n{}^m$, $\lambda^m{}_n = (i/2)\lambda_{pq}(M^{pq})^m{}_n$ where M_{mn} are the generators discussed in Chapter 1. In order to define the covariant derivative of an object like e_μ^m with mixed types of indices we need two kinds of connections, connection $\Gamma_{\mu\nu}{}^\rho$ to differentiate the world indices and connection $\omega_\mu{}^l{}_m$ to differentiate the local frame or tangent space indices. We define, to set up our notation,

$$D_\mu e_\nu^m = \partial_\mu e_\nu^m - \Gamma_{\mu\nu}{}^\alpha e_\alpha^m + \omega_\mu{}^m{}_n e_\nu^n \tag{11.4}$$

and it follows from $D_\mu(e_\lambda^m e_m^\nu) = 0$ that

$$D_\mu e_m^\nu = \partial_\mu e_m^\nu + \Gamma_{\mu\alpha}{}^\nu e_m^\alpha - \omega_\mu{}^n{}_m e_n^\nu \tag{11.5}$$

while $[\Gamma] = [\omega] = 1$. The covariant derivatives of other tensors are defined analogously. Since the local components A^m of a vector A^μ are world scalars, $\delta_{gc} A^m = \zeta^\alpha \partial_\alpha A^m$, and the covariant derivative $D_\mu A^m = \partial_\mu A^m + \omega_\mu{}^m{}_n A^n$ by definition transforms as a good tensor on all its indices, it follows that the index μ in $\omega_\mu{}^m{}_n$ is tensorial and

$$\delta_{gc}\omega_\mu{}^m{}_n = \zeta^\alpha \partial_\alpha \omega_\mu{}^m{}_n + (\partial_\mu \zeta^\nu)\omega_\nu{}^m{}_n. \tag{11.6}$$

Under the Lorentz rotations the world index is inert and we find

$$\lambda^m{}_n D_\mu A^n = \partial_\mu(\lambda^m{}_n A^n) + (\delta_L \omega_\mu{}^m{}_n) A^n + \omega_\mu{}^m{}_n \lambda^n{}_p A^p \tag{11.7}$$

† Kibble (1961); Sciama (1962), Weinberg (1972); Lévy and Deser (1978); Bergmann and Sabbata (1980.

which leads to

$$\delta_L \omega_\mu{}^m{}_n = -D_\mu \lambda^m{}_n = -\partial_\mu \lambda^m{}_n + \lambda^m{}_p \omega_\mu{}^p{}_n - \lambda^p{}_n \omega_\mu{}^m{}_p. \tag{11.8}$$

The presence of the inhomogeneous term shows that $\omega_\mu{}^m{}_n$ is a connection under Lorentz transformations. A similar discussion based on $D_\mu A^\nu = \partial_\mu A^\nu + \Gamma_{\mu\alpha}{}^\nu A^\alpha$ requires

$$D'_\mu A'^\nu(x') = \frac{\partial x'^\nu}{\partial x^\lambda} \cdot \frac{\partial x^\sigma}{\partial x'^\mu} D_\sigma A^\lambda(x)$$

and leads to

$$\delta_{gc}\Gamma_{\mu\nu}{}^\lambda = \Gamma'_{\mu\nu}{}^\lambda - \Gamma_{\mu\nu}{}^\lambda = \partial_\mu\partial_\nu\zeta^\lambda + \zeta^\alpha\partial_\alpha\Gamma_{\mu\nu}{}^\lambda + (\partial_\mu\zeta^\alpha)\Gamma_{\alpha\nu}{}^\lambda + (\partial_\nu\zeta^\alpha)\Gamma_{\mu\alpha}{}^\lambda \tag{11.9}$$

while

$$\delta_L \Gamma_{\mu\nu}{}^\lambda = 0. \tag{11.10}$$

It is not possible to recast (11.9) in a form analogous to (11.8) showing that the general coordinate transformations are distinct from the Lorentz rotations. We also note that the torsion $C_{\mu\nu}{}^\lambda = \frac{1}{2}(\Gamma_{\mu\nu}{}^\lambda - \Gamma_{\nu\mu}{}^\lambda)$ transforms like a tensor while $\delta_L(\omega_\mu{}^{mn} + \omega_\mu{}^{nm}) = 0$ as expected also from $D_\mu \eta_{lm} = -(\omega_{\mu lm} + \omega_{\mu ml})$ and $\delta_L \eta_{lm} = 0$. From the fact that $D_\mu e^m_\nu$ is a good tensor we may impose the metricity constraint $D_\mu e^m_\nu = 0$ which leads to $\Gamma_{\mu\nu}{}^\rho = e^\rho{}_m(\partial_\mu e^m_\nu + \omega_\mu{}^m{}_n e^n_\nu)$ and $D_\mu e^\nu_m = 0$ while reducing the number of independent fields. The metric tensor for the world indices is defined as the composite object $g_{\mu\nu} = \eta_{mn} e^m_\mu e^n_\nu$ and its inverse by $g^{\mu\alpha} g_{\alpha\nu} = \delta^\mu_\nu$. We then find $D_\lambda g_{\mu\nu} = -2\omega_{\lambda(mn)} e_\mu{}^m e_\nu{}^n$. The Einstein–Cartan geometry is defined by imposing the metricity postulate for $g_{\mu\nu}$, viz. $D_\lambda g_{\mu\nu} = 0$ which requires the antisymmetry in the local indices of the spinor connection, $\omega_{\lambda mn} = -\omega_{\lambda nm}$ so that $D_\lambda \eta_{mn} = 0$. The space–time connection for this geometry may be shown to take the following form:

$$\Gamma_{\mu\nu}{}^\lambda = \mathring{\Gamma}_{\mu\nu}{}^\lambda - K_{\mu\nu}{}^\lambda \tag{11.11}$$

where

$$\mathring{\Gamma}_{\mu\nu}{}^\lambda = \mathring{\Gamma}_{\nu\mu}{}^\lambda = \tfrac{1}{2} g^{\lambda\alpha}[\partial_\mu g_{\nu\alpha} + \partial_\nu g_{\mu\alpha} - \partial_\alpha g_{\mu\nu}] \tag{11.12}$$

are the symmetric Christoffel connections with respect to which the metric tensor already satisfies the metricity condition, while

$$-K_{\mu\nu}{}^\lambda = g^{\lambda\alpha}[g_{\alpha\beta} C_{\mu\nu}{}^\beta + g_{\mu\beta} C_{\alpha\nu}{}^\beta + g_{\nu\beta} C_{\alpha\mu}{}^\beta] \tag{11.13}$$

is the contorsion tensor with the symmetry $K_{\mu\nu}{}^\lambda = -K_\mu{}^\lambda{}_\nu$ and which vanishes when torsion is zero. For the Riemannian geometry with vanishing torsion the metricity condition on the tetrad gives the following spinor connection

$$\mathring{\omega}_\mu{}^{lm} = \tfrac{1}{2}[e^{l\nu}(\partial_\mu e^m_\nu - \partial_\nu e^m_\mu) + \tfrac{1}{2} e^{l\lambda} e^{m\sigma}(\partial_\sigma e_{\lambda n} - \partial_\lambda e_{\sigma n}) e^n_\mu] - (l \leftrightarrow m) \tag{11.14}$$

while for the Einstein–Cartan space–time we find

$$\omega_\mu{}^{mn} = \mathring{\omega}_\mu{}^{mn} + K_\mu{}^{mn} \tag{11.15}$$

where $K_\mu{}^{mn} = -K_\mu{}^{nm} = K_{\mu\alpha}{}^\beta e^{\alpha m} e_\beta{}^n$. We also note that

$$\mathscr{D}_\mu e_\nu^m - \mathscr{D}_\nu e_\mu^m = 2C_{\mu\nu}^\alpha e_\alpha^m \tag{11.16}$$

where

$$\mathscr{D}_\mu e_\nu^m = \partial_\mu e_\nu^m + \omega_\mu{}^m{}_n e_\nu^n = \left[\delta^m{}_n \partial_\mu + \frac{i}{2}\,\omega_{\mu pq}(M^{pq})^m{}_n\right] e_\nu^n$$

is the non-minimal covariant derivative.

The space–time curvature tensor and the Lorentz curvature tensor may be conveniently introduced by considering $[D_\rho, D_\lambda]$ acting on an arbitrary tensor A^m. When acting on the vierbein fields we find the relation (Srivastava 1983a):

$$[D_\rho, D_\lambda] e_\mu^m + 2C_{\rho\lambda}{}^\alpha D_\alpha e_\mu^m = R^\alpha{}_{\mu\rho\lambda}(\Gamma) e_\alpha^m - R^m{}_{n\rho\lambda}(\omega) e_\mu{}^n \tag{11.17}$$

where

$$\begin{aligned} -R^\mu{}_{\nu\rho\lambda}(\Gamma) &= \partial_\rho \Gamma_{\lambda\nu}{}^\mu + \Gamma_{\rho\alpha}{}^\mu \Gamma_{\lambda\nu}{}^\alpha - (\rho \leftrightarrow \lambda) \\ -R^m{}_{n\rho\lambda}(\omega) &= \partial_\rho \omega_\lambda{}^m{}_n + \omega_\rho{}^m{}_p \omega_\lambda{}^p{}_n - (\rho \leftrightarrow \lambda) \\ &= \{\partial_\rho \omega_\lambda - \partial_\lambda \omega + [\omega_\rho, \omega_\lambda]\}^m{}_n \end{aligned} \tag{11.18}$$

with $\omega_\lambda = (\omega_\lambda{}^m{}_n)$. On imposing the metricity condition for the tetrads we obtain

$$R^\mu{}_{\nu\rho\lambda}(\Gamma) = R^m{}_{n\rho\lambda}(\omega) e_\nu^n e_m^\mu. \tag{11.19}$$

The curvature scalar $R(g, \Gamma)$ may then be expressed as $R(e, \omega)$ where

$$R(e, \omega) = \tfrac{1}{2} H_{mn}^{\lambda\rho}(e) R^{mn}{}_{\rho\lambda}(\omega)$$

$$H_{mn}^{\lambda\rho}(e) = e_m^\lambda e_n^\rho - e_m^\rho e_n^\lambda = \frac{1}{2}\, e_\mu^p e_\nu^q \varepsilon_{pqmn} \left(\frac{\varepsilon^{\mu\nu\lambda\rho}}{e}\right) \tag{11.20}$$

where $e = \det(e_\mu^m) = 1/\det(e_m^\mu) = (-\det(g_{\mu\nu}))^{1/2}$ is a scalar density of weight $+1$.

The Lagrangian for the gravitational action $\sim(-g)^{1/2} R(g, \Gamma)$ in the first-order form may then be written in terms of e_μ^m and $\omega_{\lambda mn}$ as

$$\mathscr{L} = -\frac{1}{2\varkappa^2}\, eR(e, \omega) = \frac{1}{8\varkappa^2}\, e_\mu^p e_\nu^p \varepsilon_{pqmn}\, \varepsilon^{\mu\nu\lambda\rho} R^{mn}{}_{\lambda\rho}(\omega) \tag{11.21}$$

where $M = 1/\varkappa = M_{\rm Pl}/(8\pi)^{1/2} = 2.4 \times 10^{18}$ GeV. The equations of motion are obtained through Palatini's procedure by varying e_μ^m and $\omega_{\lambda mn}$

independently. The variation† with respect to the tetrad leads to

$$e(R^n{}_\sigma - \tfrac{1}{2}e^n{}_\sigma R) = \varkappa^2 \tau^n_\sigma \tag{11.22}$$

where $R^n{}_\sigma(e, \omega) = R^{nm}{}_{\rho\sigma}(\omega)e^\rho_m$ and the energy–momentum tensor τ^m_μ/e is defined through the variation in the action for the matter I_M

$$\delta I_M = \int d^4 x \tau^m_\mu \delta e^\mu_m \tag{11.23}$$

in analogy to the definition in the case of matter with integral spins of the symmetric energy–momentum tensor $\delta I_M/\delta g_{\mu\nu} = (1/2)\tau^{\mu\nu}$. With regard to the variation of $\omega_{\lambda mn}$ we derive easily the Palatini identity

$$-\delta R^{mn}{}_{\rho\lambda} = \mathscr{D}_\rho\, \delta\omega_\lambda{}^{mn} - \mathscr{D}_\lambda\, \delta\omega_\rho{}^{mn} \tag{11.24}$$

where $\delta\omega_{\lambda mn}$ is a good tensor and if we drop the surface terms we find

$$\tfrac{1}{2}\varepsilon_{pqml}\varepsilon^{\mu\nu\lambda\rho} e^q_\nu [\mathscr{D}_\rho(\omega)e^p_\mu - \mathscr{D}_\mu(\omega)e^p_\rho] = \varkappa^2 \zeta^\lambda_{ml} \tag{11.25}$$

where the spin density ζ^λ_{ml} is defined by $\delta I_M = (1/2)\int d^4 x \zeta^\lambda_{ml}\, \delta\omega_\lambda{}^{ml}$. We notice that only the non-minimal covariant derivative appears in (11.25) and if we use the metricity postulate for the tetrad this field equation reduces to an algebraic one in view of (11.16) relating torsion with the spin density of the matter fields other than gravitation.

We remark finally that the gravitino field ψ_μ transforms as a spinor under the Lorentz rotations while as a vector under the general coordinate transformations because spinors are world scalars

$$\delta\psi_\mu = \zeta^\alpha \partial_\alpha \psi_\mu + (\delta_\mu \zeta^\nu)\psi_\nu + (i/2)\lambda_{mn}M^{mn}\psi_\mu \tag{11.26}$$

where $iM_{mn} = \frac{1}{4}[\gamma_m, \gamma_n]$ are the generators of spin-1/2 field. The action for the gravitino will be discussed in the next section.

11.2.3 Lagrangian for N = 1 supergravity‡

The simplest theory of pure supergravity may be formulated in terms of the vierbein field and a Rarita–Schwinger spin-3/2 field ψ_μ. The coupling of the latter to the gravitation, however, must be most or non-minimal in order to preserve the gauge invariance constraint $\delta\psi_\mu = \partial_\mu \alpha$ of the free Rarita–Schwinger action analogous to the most minimal coupling of the Maxwell field to gravity. The supergravity Lagrangian in the second-order

† We note

$$e\varepsilon_{\mu\nu\rho\lambda} = e^m_\mu e^n_\nu e^p_\rho e^q_\lambda \varepsilon_{mnpq},\quad (1/e)\,\varepsilon^{\mu\nu\rho\lambda} = e^\mu_m e^\nu_n e^\rho_p e^\lambda_q \varepsilon^{mnpq},\quad e\,\varepsilon_{\mu\nu\rho\lambda} e^\rho_p = e^m_\mu e^n_\nu e^q_\lambda \varepsilon_{mnpq},$$

$$(1/e)\,\varepsilon^{\mu\nu\rho\lambda} e^n_\nu = e^\mu_m e^\rho_p e^\lambda_q \varepsilon^{mnpq}.$$

‡ Freedman *et al.* (1976); Deser and Zumino (1976); see also van Nieuwenhuizen (1981).

formulation is given by

$$\mathscr{L}_{\mathrm{SG}} = -\frac{1}{2\varkappa^2} eR(e, \omega(e, \psi)) - \frac{1}{4} \varepsilon^{\mu\nu\rho\sigma} \bar{\psi}_\mu \gamma_5 \gamma_\nu (\mathscr{D}_\rho \psi_\sigma - \mathscr{D}_\sigma \psi_\rho) \tag{11.27}$$

where

$$\begin{aligned} \omega_{\mu lm} &= \mathring{\omega}_{\mu lm}(e) + K_{\mu lm}(\psi) = \omega_{\mu lm}(e, \psi) \\ K_{\mu lm} &= \frac{\mathrm{i}}{4} \varkappa^2 (\bar{\psi}_\mu \gamma_l \psi_m - \bar{\psi}_\mu \gamma_m \psi_l + \bar{\psi}_l \gamma_\mu \psi_m) \end{aligned} \tag{11.28}$$

$\bar{\psi}_\mu = \psi_\mu^\dagger \gamma^{(0)}$ where $\gamma^{(0)} = (\gamma^l)_{l=0}$ is a constant matrix and† $\mathscr{D}_\rho = (\partial_\rho + \frac{1}{2}\omega_{\rho lm}\gamma^{lm})$ indicates the non-minimal covariant derivative acting on spin-1/2 fields. The curl $\mathscr{D}_{[\rho}\psi_{\sigma]}$ is covariant like the curl $\partial_{[\mu}A_{\nu]}$. Each term of the action may be shown to be invariant under diffeomorphisms as well as under local Lorentz rotations. The minimal action considered here is the complete action for simple supergravity in four-dimensional space–time. In five and eleven dimensions one needs extra four-fermion couplings and also antisymmetric tensor fields. The supergravity action above may be shown to be invariant under the following local supersymmetry transformations:

$$\delta e^l_\mu = \frac{\varkappa}{2} \bar{\varepsilon}(x) \gamma^l \psi_\mu \qquad \delta\psi_\mu = \frac{1}{\varkappa} \mathrm{D}_\mu \varepsilon(x) \tag{11.29}$$

where $\mathrm{D}_\mu \varepsilon = (\partial_\mu + \frac{1}{2}\omega_{\mu lm}(e, \psi)\gamma^{lm})\varepsilon$. These transformation laws as well as the Lagrangian can be derived by an iterative procedure in the gravitational constant by starting with the free Lagrangian ($\varkappa = 0$). The linearised vierbein can be written as $e_{\mu l} = \eta_{\mu l} + \varkappa h_{\mu l}$ where $h_{\mu l} = h_{l\mu}$ describes the free graviton spin-2 field. For $\varkappa = 0$ the Lagrangian is invariant under two separate Abelian gauge transformations

$$\delta h_{\mu\nu} = \partial_\mu \zeta_\nu(x) + \partial_\nu \zeta_\mu(x) \qquad \delta\psi_\mu = \delta_\mu \alpha(x) \tag{11.30}$$

and the following rigid supersymmetry transformations

$$\delta h_{\mu\nu} = \bar{\varepsilon}\gamma_\mu \psi_\nu + \bar{\varepsilon}\gamma_\nu \psi_\mu \qquad \delta\psi_\mu = \tfrac{1}{2}\mathring{\omega}_{\mu lm}\gamma^{lm}\varepsilon. \tag{11.31}$$

We may use Noether's step by step procedure mentioned before to arrive at the nonlinear Lagrangian along with the local supersymmetry transformation laws. We note that at the coupled level the Abelian transformations independent at the linearised level combine into an irreducible non-Abelian local SS transformation law. This is analogous to the case of the local Yang–Mills transformation $\delta v^a_l = \mathrm{D}_l \Lambda^a = \partial_l \Lambda^a + f^a{}_{bc} v_l{}^b \Lambda^c$ which at the linearised level also splits into an Abelian gauge transformation and a global Yang–Mills rotation.

The local algebra, i.e. the commutator of two local symmetries, may be calculated straightforwardly. We find, for example, that as in the case of

† $\gamma^{lm} = \frac{1}{4}[\gamma^l, \gamma^m] = \mathrm{i}M^{lm}$.

rigid supersymmetry in the local case as well the commutator $[\delta_s, \delta_s]$ when acting on the fermionic field ψ_μ generates a term proportional to the gravitino field equations indicating the need for introducing auxiliary fields in the theory which are known for simple supergravity in the second-order formulation. Moreover, along with producing the expected general coordinate transformation there are also found terms on the RHS which represent a local Lorentz and a local supersymmetry transformation. We will not dwell on these details and remark only that including auxiliary fields, a scalar, a pseudoscalar and an axial vector A^μ, the local algebra in the second-order formulation does 'close' contrary to the case of the first-order formulation. The supergravity theory may also be shown to be a gauge theory of the super-Poincaré group and may also be derived as the geometry of superspace (van Nieuwenhuizen 1981, Ne'eman and Regge 1978, Wess and Bagger 1983, Gates *et al.* 1983, Fradkin and Tseytlin 1985). Corresponding to the graded conformal group a conformal supergravity theory can also be constructed.

11.3 Supergravity coupling to matter

11.3.1 Non-linear realisation of supersymmetry. Coupling of Volkov–Akulov field

When global supersymmetry is realised in the spontaneously broken mode the resulting 'goldstino' field λ corresponding to the broken $N=1$ supersymmetry has a nonlinear SS transformation law along with an inhomogeneous term given by (Volkov and Akulov 1973, Samuel and Wess 1983, Wess 1983):

$$\delta\lambda = d\varepsilon + \frac{i}{d}(\bar{\lambda}\gamma^l\varepsilon)\partial_l\lambda \tag{11.32}$$

where ε and λ are Majorana spinors and we use four-component notation. The unbroken Poincaré transformations are, however, realised linearly on λ. This is analogous to the case of the original sigma model where the massless Goldstone pions transform nonlinearly under the broken SU(2) generators while linearly under the unbroken SU(2) and the effective low-energy Lagrangian is a nonlinear model. The constant 'd' of dimension 2 indicates the square of the SS breaking scale and $\langle\delta\lambda\rangle_0 = d\varepsilon \neq 0$ indicates that the supersymmetry is broken. It is easily shown that the above nonlinear transformation closes into the supersymmetry algebra

$$[\delta_2, \delta_1]\lambda = -2i(\bar{\varepsilon}_2\gamma^l\varepsilon_1)\partial_l\lambda. \tag{11.33}$$

No other fields are needed to make the realisation faithful. We note also that if $\rho(x)$ is another field with the homogeneous transformation law

$$\delta\rho = (\mathrm{i}/d)(\bar{\lambda}\gamma^l\varepsilon)\partial_l\rho(x)$$

then the algebra closes on ρ as well. The simplest nonlinear Lagrangian invariant under supersymmetry up to a divergence is given by

$$\begin{aligned}\mathscr{L}_\lambda &= -\frac{d^2}{2}\det(\delta^l_m - \frac{\mathrm{i}}{d^2}\bar{\lambda}\gamma^l\partial_m\lambda) \\ &= -\frac{d^2}{2} + \frac{\mathrm{i}}{2}\bar{\lambda}\gamma^l\partial_l\lambda + \ldots + \mathrm{O}(d^{-6})\end{aligned} \tag{11.34}$$

where the dots represent the interaction terms $\frac{1}{2}(T^m{}_m T^n{}_n - T_m{}^n T_n{}^m) + \mathrm{O}(T^3) + \mathrm{O}(T^4)$ where $T_m{}^n = (\mathrm{i}/d^2)\bar{\lambda}\gamma_m\partial^n\lambda$ and the series terminates since $\lambda(\bar{\lambda}\lambda)^2 = 0$. The Noether supercurrent is derived to be

$$J_m = \mathrm{i}d\gamma_m\lambda + \ldots \qquad \partial_m J^m \overset{\circ}{=} 0 \tag{11.35}$$

and $\langle 0 |\, J^l_A \,| \lambda_B\rangle = \mathrm{i}d(\gamma^l)_{AB}$. The positive vacuum energy density $d^2/2$ shows again that the supersymmetry is broken.

When the supersymmetry is promoted to be a local one a non-diagonal term $\varkappa\bar{\psi}_l J^l_N = \varkappa d(\bar{\psi}\gamma\lambda)$ will be added to the Lagrangian, say, if we follow the Noether coupling procedure. Deser and Zumino (1977) studied the coupling of the Volkov–Akulov Lagrangian to de Sitter supergravity† and showed that under the assumption of a vanishing cosmological term, the gravitino acquires a mass $\sim d^2$ while the goldstino field can be absorbed into a redefinition of the fields e^m_μ and ψ_μ. The super-Higgs effect for general interaction of the chiral supermultiplets and gauge supermultiplet will be described in the following sections. We remark that the Higgs mechanism for gauged supersymmetry algebra is distinct from the same mechanism for the gauged Lie algebras owing to certain positivity properties of superalgebras (Witten 1981, 1982). Supersymmetry restricts the scalar field content which determines the spontaneous breaking and the vacuum energy. The overall scalar potential has a unique form once the supersymmetry variations of the fermionic fields in the theory are given.

† The V–A Lagrangian gives rise to a negative cosmological term $-d^2e/2$. We may add to the Lagrangian a supersymmetric cosmological term $3m^2e + \mathrm{i}me\bar{\psi}_\mu\sigma^{\mu\nu}\psi_\nu$ with the positive cosmological constant while slightly modifying the transformation laws to have a local SS invariance. The vanishing of the cosmological constant in the total Lagrangian determines $m \sim d$. Choosing the gauge $\lambda = 0$ (or alternatively redefining the gravitino field through a local SS transformation with a parameter λ which eliminates the propagation of the goldstino field) we obtain in the theory besides gravitation a massive gravitino apart from the interaction terms.

11.3.2 General coupling to Yang–Mills theory†

The $N=1$ pure supergravity Lagrangian can be coupled in a locally supersymmetric fashion to an arbitrary Yang–Mills system which is specified by the gauge group K and by the transformation properties under K of the set of chiral matter supermultiplets. The presence of the dimensional coupling constant $\varkappa$ in the pure supergravity Lagrangian leads to a non-renormalisable theory even if the matter–gauge system coupled to it is renormalisable one. We should rather demand instead that after the coupling to supergravity the resulting non-renormalisable terms are such that in the flat space limit, $M_{\mathrm{Pl}} \rightarrow \infty$, the theory becomes renormalisable. In supergravity a spontaneously broken, locally supersymmetric theory admits as a global limit an explicitly broken, supersymmetric theory with soft SS breaking terms. The most complete form of the interacting theory was given by Cremmer *et al.* and we will adopt their notation. The component fields of left-handed chiral supermultiplets S_i transforming according to representation R of the gauge group K will be indicated by $(z_i, \chi_{\mathrm{L}i}, h_i)$ where i labels the representation index, $T_i^{\alpha j}$ stands for the generators in the representation and α is the group index labelling the adjoint representation. The gauge fermions are called λ^α while $F^\alpha_{\mu\nu}$ indicates the field strengths of the gauge bosons.

The arbitrariness of the interacting supergravity–Yang–Mills Lagrangian consists essentially in a non-canonical modification of the chiral and Yang–Mills kinetic terms. They involve respectively a real gauge invariant function $G(z, z^*)$, called the Kähler potential, and an analytic (chiral) function of the complex scalar fields z_i written $f_{\alpha\beta}(z)$ which transforms as the symmetric product of the adjoint representation of K. The requirement of local supersymmetry in its turn proliferates them in the interaction terms as well as in the local SS transformation laws. The scalar field in supergravity theories appears as coordinates of a Kähler manifold whose metric enters in the scalar kinetic terms which has a form characteristic of the supersymmetric nonlinear sigma model. The final Lagrangian has the following form:

$$\mathscr{L} = \mathscr{L}_{\mathrm{BK}} + \mathscr{L}_{\mathrm{P}} + \mathscr{L}_{\mathrm{FK}} + \mathscr{L}_{\mathrm{FM}} + \mathscr{L}_{(4)\mathrm{F}} \tag{11.36}$$

where the bosonic part is given by ($\varkappa = 1$)

$$e^{-1}\mathscr{L}_{\mathrm{BK}} = -\frac{1}{2}R - G^i{}_j \mathrm{D}_\mu z_i \mathrm{D}^\mu z^{*j} - \frac{1}{4}\mathrm{Re}\, f_{\alpha\beta} F^\alpha_{\mu\nu} F^{\mu\nu\beta} + \frac{\mathrm{i}}{4}\mathrm{Im}\, f_{\alpha\beta} F^\alpha_{\mu\nu} \tilde{F}^{\mu\nu\beta}$$

$$-e^{-1}\mathscr{L}_{\mathrm{P}} = V(z, z^*) = e^G[G^i (G^{-1})_i{}^j G_j - 3] + \frac{1}{2}\mathrm{Re}(f^{-1}_{\alpha\beta}\mathrm{D}^\alpha \mathrm{D}^\beta) = V_{\mathrm{c}} + V_{\mathrm{g}} \tag{11.37}$$

† Cremmer *et al* (1978, 1979, 1982, 1983a), Chamseddine *et al.* (1982), Witten and Bagger (1982), Bagger (1983).

where V_c and V_g are the chiral and gauge parts of the scalar potential respectively. The covariant derivatives are covariant with respect to the gravity and the gauge group and we define $G^i = \partial G/\partial z_i$, $G_i = \partial G/\partial z^{*i}$, $G^i{}_j = \partial^2 G/\partial z_i \partial z^{*j}$, $D_\alpha = g_\alpha G^i (T_\alpha)_i{}^j z_j$ with g_α indicating the gauge coupling constant associated to the normalised generators. We note that in the positivity domain of the spin-1 kinetic term the gauge field contribution V_g is a semi-positive definite function. We will describe later the necessary and sufficient conditions to obtain a semi-positive definite V_c. The fermionic kinetic and mass terms are listed below while for the $\mathscr{L}_{(4)F}$ we refer the reader to the original reference where the SS transformation laws are also given in complete form†

$$
\begin{aligned}
e^{-1}\mathscr{L}_{FK} = & -\frac{1}{4}\,\mathrm{Re}\, f_{\alpha\beta}\bar{\lambda}^\alpha \not{D}\lambda^\beta - \frac{1}{4}e^{-1}\varepsilon^{\mu\nu\rho\sigma}\bar{\psi}_\mu\gamma_5\gamma_\nu D_\rho\psi_\sigma \\
& - G^i{}_j\bar{\chi}_{Li}\not{D}\chi_R{}^j - \frac{i}{8}\,\mathrm{Im}\, f_{\alpha\beta}e^{-1}D_\mu(e\bar{\lambda}^\alpha\gamma_5\gamma^\mu\lambda^\beta) + \mathrm{HC} \\
e^{-1}\mathscr{L}_{FM} = & \; e^{G/2}\bar{\psi}_{\mu R}\sigma^{\mu\nu}\psi_{\nu R} + \bar{\psi}_R\cdot\gamma\left\{e^{G/2}G^i\chi_{Li} - \frac{i}{2}\tilde{g}G^i T_i^{\alpha j}z_j\lambda_L^\alpha\right\} \\
& - e^{G/2}\bar{\chi}_{Li}[G^{ij} + G^i G^j - G^l G_l{}^{-1k}G_k^{ij}]\chi_{Lj} + 2i\tilde{g}G_i^k z^{*j}T_j^{\alpha i}\bar{\lambda}_L^\alpha\chi_{Lk} \\
& + \frac{1}{2}f^k_{\alpha\beta}\left\{\frac{1}{2}e^{G/2}G_l G_k^{-1l}\bar{\lambda}_L^\alpha - i\tilde{g}\,\mathrm{Re}\, f^{-1}_{\alpha\gamma}G^i T_i^{\gamma j}z_j\bar{\chi}_{Lk}\right\}\lambda_L^\beta + \mathrm{HC}
\end{aligned}
\tag{11.38}
$$

where D_μ is covariant both with respect to the gauge group and gravitation. We list also the scalar field contributions to the fermionic local SS variations of the fermion fields:

$$
\begin{aligned}
\delta\lambda_R^\alpha &= +\frac{i}{2}\varepsilon_R \mathrm{Re} f^{-1}_{\alpha\beta}D_\beta + \ldots \\
\delta\chi_{Li} &= -\frac{1}{2}\varepsilon_L e^{G/2}(G^{-1})_i{}^j G_j + \ldots \\
\delta\psi_\mu &= \frac{1}{2}\gamma_\mu\varepsilon_R e^{G/2} + \ldots .
\end{aligned}
\tag{11.39}
$$

We remark that in arriving at the above Lagrangian it is necessary that G has the following form‡

$$G(z, z^*) = J(z, z^*) + \ln|g(z)|^2 \tag{11.40}$$

where the superpotential $g(z)$ is a non-vanishing analytic function of z_i. We

† See also Ferrara (1984) and Nath (1984), where a complete list of references may be found.

‡ When $g = 0$ proper substitution rules must be used to obtain the correct Lagrangian.

note that J and g are defined up to a Kähler transformation

$$J \to J + f(z) + f^*(z^*)$$
$$g(z) \to g(z)e^{-f(z)}. \tag{11.41}$$

The Lagrangian above is also implied in a superfield formulation where the extension of the global supersymmetry action to local supersymmetry should look like

$$\int d^8zE\left[\Phi(S, \bar{S}e^{2V}) + \mathrm{Re}\left(\frac{1}{R}\,g(S)\right) + \mathrm{Re}\left(\frac{1}{R}\,f_{\alpha\beta}(S)W^{\alpha}_{a}\varepsilon^{ab}W^{\beta}_{b}\right)\right] \tag{11.42}$$

where Φ, g, f are three input functions, R is the chiral scalar curvature superfield and E is the superspace determinant. In the complete Lagrangian Φ and g lose their independent meaning and enter only through G above, where J is related to Φ. The goldstino field is uniquely identified in the local supersymmetry under consideration by the spin-1/2 fermion which couples to the gravitino gauge field in $\mathscr{L}_{\mathrm{FM}}$

$$\eta_{\mathrm{L}} = -\left[e^{G/2}G^i\chi_{\mathrm{L}i} - (i/2)D^{\alpha}\lambda^{\alpha}_{\mathrm{L}}\right]. \tag{11.43}$$

11.3.3 The super-Higgs effect in N = 1 supergravity

In the standard Higgs mechanism the local gauge invariance allows a spin-0 Goldstone boson, corresponding to a spontaneously broken rigid symmetry, to be rotated away such that the initially massless gauge boson with states of helicity ± 1 acquires a third helicity-0 state becoming consequently massive—the spontaneous breaking of a local gauge symmetry. The corresponding super-Higgs mechanism may be shown to occur and the problem of the apparent non-existence of the massless goldstino particle of spontaneously broken rigid supersymmetry is thus overcome in the context of local supersymmetry.

We observed earlier that V_g is semi-positive definite in the positivity domain of the kinetic term of the Yang–Mills field. The chiral part V_c of the scalar potential is the difference between two positive definite terms due to the positivity properties of the Kähler metric $G^i{}_j$ which is present in the kinetic terms of the spin-0 and spin-1/2 particles of the chiral supermultiplets and as such may assume positive, negative or vanishing value. Thus we may obtain the super-Higgs effect with vanishing vacuum energy (Minkowski space) since it is no longer an order parameter for broken local supersymmetry contrary to the case of rigid supersymmetry. Moreover broken supergravity is possible even in the presence of a single chiral supermultiplet.

Through the minimal coupling $\bar{\psi}_{\mathrm{R}}\gamma\eta_{\mathrm{L}} + \mathrm{HC}$ the local supersymmetry allows the goldstino to be rotated away by a special choice of the supersymmetric gauge while the previously massless gravitino with helicity states $\pm 3/2$ acquires helicity states $\pm 1/2$ and becomes massive. A necessary and sufficient condition for spontaneously broken supersymmetry requires that one of the quantities

$$e^{G/2}G^i \qquad D^\alpha = g_\alpha G^i (T^\alpha)_i{}^j z_j \tag{11.44}$$

is different from zero at the minimum of the scalar potential, i.e. $(\partial V/\partial z_i)_{z=z_0} = 0$. If we arrange also a vanishing energy at the minimum, $V_0 = V(z_0, z_0^*) = 0$, i.e. a vanishing cosmological constant, then Minkowski space is a solution of the vacuum field equations and on this background the gravitino mass has its usual meaning and it is given by the value of the Kähler potential $m_{3/2} = Me^{G/2}$ analogously to the manner in which the value $\langle H \rangle_0$ fixes the gauge boson mass.

In the absence of supersymmetry breaking, $G^i = 0$, $D^\alpha = 0$, the vacuum energy, $V_0 = -3M^4 e^G < 0$, is negative which corresponds to anti-de Sitter ($G \neq 0$) or Poincaré ($G = -\infty$) supergravity with multiplets degenerate in mass. The super-Higgs effect due to the broken supersymmetry on the other hand may occur in Minkowski, de Sitter or anti-de Sitter space and $V_0 > 0$ always describes a broken supergravity.

We consider first the case of 'minimal' coupling of the supergravity to a Yang–Mills system defined by

$$G^i{}_j = \delta^i_j \qquad f_{\alpha\beta} = \delta_{\alpha\beta} \tag{11.45}$$

so that all the kinetic terms are canonical and the Lagrangian depends only on the superpotential $g(z)$. We have

$$\begin{aligned} G(z, z^*) &= z_i z^{*i}/M^2 + \ln | g(z) |^2/M^6 \\ G^i &= M\partial G/\partial z_i = \frac{M}{g}\left(g^i + \frac{z^{*i}}{M^2} g\right) \\ m_{3/2} &= Me^{G/2} = e^{zz^*/2M^2}(| g(z) | /M^2) \\ V_c &= e^{zz^*/M^2}[\, |D_{z_i} g|^2 - 3(| g(z) |^2/M^2)] \end{aligned} \tag{11.46}$$

where $g^i = \partial g/\partial z_i$ and $D_{z_i} = \partial/\partial z_i + z^{*i}/M^2$ is the Kähler covariant derivative. The supersymmetry is broken if at least one of the Kähler covariant derivatives of the superpotential is non-vanishing. When $M \to \infty$ with z_i fixed gravity becomes unimportant leaving behind the familiar condition of broken global supersymmetry.

In the simplest case of a constant superpotential, $g = m^3$, ($D^\alpha = 0$), we find

$$V = m^6 e^{z_i z^{*i}/M^2}(zz^*/M^4 - 3/M^2) \tag{11.47}$$

which has its stationary points at $z_i = 0$ and $z = \pm\sqrt{2}M$ (figure 11.1). The supersymmetry is unbroken for $z_{i0} = 0$ with $V_0 = -3M^2\ (m_{3/2})^2$. This point, however, corresponds to a local maximum of the potential. The other values on the contrary correspond to minima with some non-vanishing $D_z g = z^*/M^2$. The SS is broken with a negative vacuum energy $V_0 = -e^2 m^6/M^2$ corresponding to an anti-de Sitter vacuum with the 'apparent' gravitino mass em^3/M^2. This example shows that supersymmetry may now be broken even in the presence of one matter supermultiplet and the supergravity breaking minima may be the lowest ones even if supersymmetric stationary points exist.

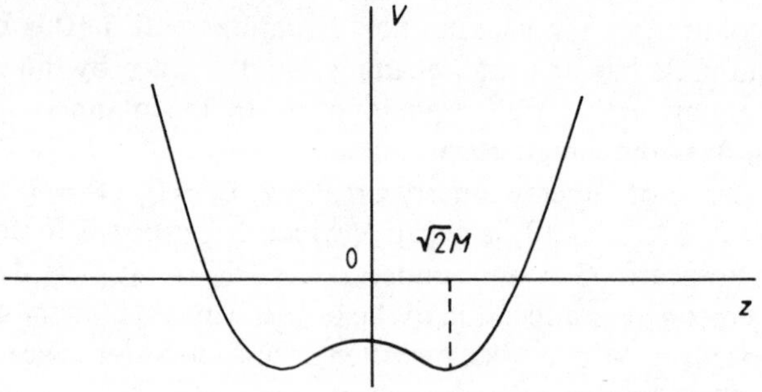

Figure 11.1 Scalar potential with constant g for 'minimal' coupling.

The Polonyi superpotential $g = m^2(z+\beta)$ offers itself as an example where we may fine tune the parameter β such that a broken supersymmetry minimum is obtained with vanishing vacuum energy, $V_{co} = 0$. We find $D_z g = m^2\ [1 + z^*(z+\beta)/M^2]$ and $D_z g = 0$ does not have any solution for z in the case $|\beta| < 2M$ and consequently the supersymmetry is certainly broken for such values of β. The chiral potential is given by

$$V_c = \left(\frac{m}{M}\right)^4 e^{|z|^2/M^2}\,[\,|M^2 + z^*(z+\beta)|^2 - 3M^2|z+\beta|^2]. \quad (11.48)$$

It is straightforward to show that we can obtain an absolute minimum with $V_c = 0$ at $z_0 = \pm(\sqrt{3}-1)M$ if $\beta = \pm(2-\sqrt{3})M$ and where $D_z g = \sqrt{3}m^2$ with the gravitino mass given by

$$m_{3/2} = (m^2/M)\, e^{(\sqrt{3}-1)^2/2}. \quad (11.49)$$

Performing the shift, say, $z \to z + (\sqrt{3}-1)M$ in V_c we may calculate the mass squared of the real scalars which are found to be $2\sqrt{3}(m_{3/2})^2$ and $2(2-\sqrt{3})\,(m_{3/2})^2$. The scale of the supersymmetry breaking defined by the first term of (11.46) is given by:

$$M_s^2 = \langle e^{zz^*/2M^2}\,|D_z g|\rangle_0 = \sqrt{3}m^2\, e^{(\sqrt{3}-1)^2/2} \quad (11.50)$$

and we find $M_s^2 = \sqrt{3}Mm_{3/2}$. We note that the scalar field acquires a VEV of the order of the Planck mass while the scalar masses are of the order of the gravitino mass and

$$m_{3/2} = \frac{M_s^2}{M_{Pl}} \sqrt{\frac{8\pi}{3}}. \tag{11.51}$$

For the minimal coupling the term $\mathscr{L}_{FM}$ which determines the fermion mass matrix reduces to

$$e^{-1}\mathscr{L}_{FM} = e^{G/2}[\bar{\psi}_{\mu R}\sigma^{\mu\nu}\psi_{\nu R} - \bar{\psi}_R\gamma\tilde{\eta}_L - \tfrac{2}{3}\bar{\tilde{\eta}}_L\tilde{\eta}_L]$$
$$+ \bar{\chi}_{Li}M^{ij}\chi_{Lj} + 2\bar{\chi}_{Li}M^{i\alpha}\lambda_L^\alpha + \bar{\lambda}_L^\alpha M^{\alpha\beta}\lambda_L^\beta + \text{HC} \tag{11.52}$$

where $\tilde{\eta}_L = e^{-G/2}\eta_L$ and the spin-1/2 fermion matrix has the form

$$M^{ij} = -e^{G/2}[G^{ij} + \tfrac{1}{3}G^iG^j]$$
$$M^{i\alpha} = -(i/3)G^i D^\alpha + iD^{\alpha i} \tag{11.53}$$
$$M^{\alpha\beta} = -\tfrac{1}{6}e^{-G/2}D^\alpha D^\beta.$$

The quadratic mass relation may then be derived to be

$$\text{Str}\, M^2 = \sum_{J=0}^{3/2}(-1)^{2J}(2J+1)m_J^2 = (N-1)(2m_{3/2}^2 - \varkappa^2 D^\alpha D^\alpha)$$
$$-2\tilde{g}_\alpha D^\alpha \text{Tr}\, T^\alpha \tag{11.54}$$

where the last term is only possible for Abelian U(1) factors of K with Tr $T^\alpha \neq 0$. If we set $m_{3/2} = 0$ and $\varkappa = 0$ we get back the mass relations of spontaneously broken globally supersymmetric Yang–Mills theories. The above relation is a consequence of the simultaneous occurrence of the Higgs and super-Higgs effects.

It has recently been realised that certain $N=1$ supergravity model theories may have semi-positive definite potentials permitting them at the tree level a vanishing cosmological constant even in the presence of broken supersymmetry. A necessary and sufficient condition (Barbieri *et al.* 1985) for having a semi-positive definite potential in the general case is given by $\det(-\varphi) \leqslant 0$ where φ is the Hermitian matrix $(\partial^2/\partial z_i \partial z^{*j})\exp(-G/3)$. In particular the necessary and sufficient condition for a flat potential ($V_c \equiv 0$) is $\det(\varphi) = 0$.

The chiral potential may be rewritten as follows, N denoting the number of chiral supermultiplets:

$$V_c = \frac{9}{N^2}e^{(N+3)/3}(G^{-1})_i{}^j\partial^i\partial_j e^{-(N/3)G}. \tag{11.55}$$

The flatness of the potential requires a particular Kähler potential such that

$$(G^{-1})_i{}^j\partial^i\partial_j e^{-(N/3)G} = 0. \tag{11.56}$$

A particular solution is

$$G = -\frac{3}{N}\sum_{i=1}^{N}\ln[f_i(z) + f_i^*(z^*)] \tag{11.57}$$

which is equivalent to

$$G = -\frac{3}{N}\sum_{i=1}^{N} \ln(z_i + z^{*i})$$

up to field redefinitions $z \to f(z)$. The curvature tensor for the Kähler manifold $R^i{}_j$ is defined by†

$$R^i{}_j = \partial^i \partial_j \ln \det(G^i{}_j). \tag{11.58}$$

For the above solution we obtain

$$R^i{}_j = \frac{2N}{3} G^i{}_j \tag{11.59}$$

indicating an Einstein manifold. However, this property alone is not enough to ensure the flatness of the potential. We remark that for flat potentials the vanishing of the cosmological constant occurs naturally at the classical level whether the supersymmetry is broken or not. The gravitino mass in such models is undetermined and in such so-called 'no-scale' models one hopes to get all the low energy scale parameters from the Planck mass through the radiative corrections and the mechanism of dimensional transmutation (and the renormalisation group equation). One of the main features of such supergravity models is the non-minimality of the kinetic terms of the scalars which form a non-compact symmetric Kähler manifold, viz. SU(1,1)/U(1) in the less symmetric case and SU(n,1)/SU(n) × U(1) in the maximally symmetric case where n here is the number of gauge non-singlet complex scalar fields. Such global non-compact groups play an essential role in $N \geqslant 4$ extended supergravity theories. The non-compact group invariance seemingly guaranteeing a flat potential may be a relic of an underlying bigger theory. In fact the effective potential derived from the $E_8 \times E_8$ superstring model (Green and Schwartz 1984, Gross *et al.* 1985) also has such a non-compact symmetry.

Consider the case of one scalar field with ($\varkappa = 1$)

$$G(z, z^*) = -3 \ln(z + z^*) + \ln |c|^2 \tag{11.60}$$

where $g = c$ is a constant superpotential. We find $e^{G/2}G_z = -3|c|/(z+z^*)^{5/2}$, $m_{3/2} = e^{G/2} = |c|/(z+z^*)^{3/2}$, $R_{zz^*} = (2/3)G_{zz^*}$ and $V_c \equiv 0$ for any value of z. The supersymmetry is broken when $c \neq 0$ but the gravitino mass is non-vanishing but undetermined. The Kähler manifold is an Einstein space with constant curvature and its isometries form a non-compact SU(1,1) group. The Lagrangian for the gauge singlet scalar z reads as

$$3\sqrt{-g}\,\frac{(\partial_\mu z)(\partial^\nu z^*)}{(z+z^*)^2}\, g^{\mu\nu} \tag{11.61}$$

† $R^i{}_j = 0$ for the flat Kähler manifold, $G^i{}_j = \delta^i{}_j$. A generalisation to the locally flat case is given in Grisaru *et al.* (1982) and Gates *et al.* (1983).

which describes a nonlinear sigma model with an SU(1,1)/U(1) global symmetry. Furthermore the SU(1,1) Mobius transformations

$$z \to \left(\frac{\alpha z + \mathrm{i}\beta}{\mathrm{i}\gamma z + \delta}\right) \qquad \alpha, \beta, \gamma, \delta \text{ real with } \alpha\delta + \beta\gamma = 1 \tag{11.62}$$

leave the whole Lagrangian, except the gravitino–goldstino mass term, invariant after simultaneous chiral rotations on the fermionic fields. For $c = 0$ the supersymmetry is not broken and $m_{3/2} = 0$ while all SU(1,1) breaking terms drop out from the Lagrangian. For $c \neq 0$ the supersymmetry is spontaneously broken and simultaneously the SU(1,1) symmetry is broken down to an U(1) defined by the imaginary translations $z \to z + \mathrm{i}\beta$.

The existence of non-compact and anomaly-free global symmetry seems necessary along with a 'no-scale' model to obtain a vanishing cosmological constant (Cremmer *et al.* 1983b, Ellis *et al.* 1984, Antoniadis *et al.* 1985). The SU(n, 1) 'no-scale' model is based on the following G:

$$G = -3\ln\left(z + z^* - \frac{\varphi_i \varphi^{*i}}{3}\right) + \ln|g(\varphi_i)|^2 \tag{11.63}$$

where the superpotential g depends only on the 'observable' (sector) gauge non-singlet chiral superfields φ_i while the singlet z belongs to the 'hidden' sector and has the form

$$g(\varphi_i) = c + d_{ijk}\varphi^i\varphi^j\varphi^k. \tag{11.64}$$

The gauge singlet plays a similar role to the simple case considered above and the parameter c breaks supersymmetry spontaneously. The scalar potential is positive semi-definite and flat along the directions $(3/2)^{1/2}$ $\mathrm{i}(z - z^*)$ and $-(3/\sqrt{6})\ln(z + z^* - \varphi_i\varphi^{*i}/3)$.

Recently, superstring models have been proposed to solve the problems of quantum gravity, of unifying all the fundamental interactions, of flavour and of the cosmological constant. The effective low energy theory obtained from the ten-dimensional $E_8 \times E_8$ superstring theory after the compactification of the extra six dimensions on a Ricci flat manifold is found to be based on the following Kähler potential (Witten 1985, Dine *et al.* 1985, Cohen *et al.* 1985):

$$G = -3\ln(z + z^* - \varphi_i\varphi^{*i}) + \ln|W(S, \varphi_i)|^2 - \ln(S + S^*) \tag{11.65}$$

and a simple non-trivial chiral kinetic function for the gauge superfield strength W^α, $f_{\alpha\beta}W^\alpha W^\beta$, $f_{\alpha\beta} = \delta_{\alpha\beta}S$. Here $W(S, \varphi_i)$ is an effective S-dependent superpotential of another 'hidden' sector gauge singlet chiral superfield S which is generated by the 'hidden E_8' gaugino condensation and the gauge non-singlet observable fields φ_i. The resulting theory has a 'no-scale' (SU(1,1)/U(1)) × (SU(n, 1)/(SU(n) × U(1)) structure. The local SS breaking scale at the tree level is a non-vanishing (undetermined) gravitino mass, which has to be fixed by, say, a dynamical determination of S. The limiting low energy theory† is obtained by taking the flat limit, $M_{\rm Pl} \to \infty$ with $m_{3/2}$ fixed and dropping out decoupled and superheavy fields. The renormalised non-minimal kinetic function for the gauge field, $f_{\alpha\beta} = (S + b \ln {\rm U}(\varphi_i))$, gives rise to a gaugino mass $m_{1/2}$ and it is argued currently that it is responsible for the dominant source of global supersymmetry breaking in the observable sector to produce via radiative corrections non-zero scalar masses (for sleptons, for example) for the gauge non-singlet fields φ_i. The supergravity theories may then be regarded as effective theories for the light particle states once integration over the infinitely many massive modes of the superstring spectrum has been performed.

Problem

11.1 Discuss the limiting low energy theory when the superpotential is of the form

$$g(z, y) = h(z) + g(y)$$

where the scalar fields z belong to the 'hidden' sector while y belongs to the 'observable' sector. The fields in the hidden sector assume a vacuum expectation value of the order of the Planck mass, while the observable sector does not contain this large scale and the vacuum expectation values are much smaller than the Planck mass. See Nilles (1984).

† For earlier attempts to build superunified models see the review of Nilles (1984) and Ellis (1985).

REFERENCES

Albert A A 1948 *Trans. Am. Math. Soc.* **64** 552

Amati D and Chou K 1982 *Phys. Lett.* **114B** 129

Antoniadis I, Kounnas C and Nanopoulos D V 1985 *CERN preprint TH4187/85*

Arnowitt R, Nath P and Zumino B 1975 *Phys. Lett.* **56B** 81

Bagger J 1983 *Nucl. Phys.* B **211** 302

Bailin D and Love A 1986 *Introduction to Gauge Field Theory* (Bristol: Hilger)

Barbieri R, Cremmer E and Ferrara S 1985 *CERN preprint TH4177/85*

Belinfante F J 1940 *Physica* **7** 449

Berezin F A 1966 *The Method of Second Quantization* (New York: Academic)

Bergmann P G and Sabbata V (ed) 1980 *Cosmology and Gravitation* (New York: Gordon and Breach)

Callan C G, Coleman S and Jackiw R 1970 *Ann. Phys., NY* **59** 42

Chamseddine A, Arnowitt R and Nath P 1982 *Phys. Rev. Lett.* **49** 970

Cohen E, Ellis J, Gomez C and Nanopoulos D V 1985 *CERN preprint TH4159/85*

Coleman S and Mandula J 1967 *Phys. Rev.* **159** 1251

Cremmer E *et al.* 1978 *Phys. Lett.* **79B** 231

—— 1979 *Nucl. Phys.* B **147** 105

—— 1982 *Phys. Lett.* **116B** 231

—— 1983a *Nucl. Phys.* B **212** 413

—— 1983b *Phys. Lett.* **133B** 61

Deser S and Zumino B 1976 *Phys. Lett.* **62B** 335

—— 1977 *Phys. Rev. Lett.* **38** 1433

Dine M *et al.* 1985 *Princeton preprint*

Ellis J 1984 *CERN preprint TH4017/84*

—— 1985 in *Relativity, Supersymmetry and Cosmology: Lectures at SILARG V, Bariloche* (Singapore: World Scientific)

Ellis J *et al.* 1984 *Nucl. Phys.* B **241** 406; B **247** 373

Fayet P 1976 *Nuovo Cimento* A **31** 626

Fayet P and Ferrara S 1977 *Phys. Rep.* **32C** 249†

Fayet P and Illiopoulos J 1974 *Phys. Lett.* **51B** 461

Ferrara S 1984 in *Supersymmetry and Supergravity/Nonperturbative QCD, Lecture Notes in Physics* no 208 (Berlin: Springer)

Ferrara S and Piguet O 1975 *Nucl. Phys.* B **93** 261

Ferrara S and Remiddi E 1974 *Phys. Lett.* **53B** 347

Ferrara S, Savoy C A and Zumino B 1981 *Phys. Lett.* **100B** 393

Ferrara S, Wess J and Zumino B 1974 *Phys. Lett.* **51B** 239

Ferrara S and Zumino B 1974 *Nucl. Phys.* B **79** 413

—— 1975 *Nucl. Phys.* B **87** 207

Fischler W, Nilles H P, Polchinski J, Raby S and Susskind L 1983 *Preprint SLAC PUB-2760*

Fradkin E S and Tseytlin A A 1985 *Phys. Rep.* **119** 233
Freedman D, van Nieuwenhuizen P and Ferrara S 1976 *Phys. Rev.* D **13** 3214
Gates S J, Grisaru M T, Roček M and Siegel W 1983 *Superspace* (Reading, Mass.; Benjamin–Cummings)
Gol'fand Yu A and Likhtman E P 1971 *JETP Lett.* **13** 323
Green M B and Schwartz J 1984 *Phys. Lett.* **149B** 117
Grisaru M T, Roček M and Karlhede A 1982 *Phys. Lett.* **120B** 189
Grisaru M T, Roček M and Siegel W 1979 *Nucl. Phys.* B **159** 429
Gross D, Harvey J A, Martinec E and Rohm R 1985 *Phys. Rev. Lett.* **54** 502
Haag R, Lopszanski J T and Sohnius M F 1975 *Nucl. Phys.* B **88** 257
Heine V 1957 *Phys. Rev.* **107** 620
't Hooft G and Veltman M 1973 *Diagramar CERN (Yellow) Rep. 73–9*
Howe P S, Stelle K S and Townsend P K 1981 *Nucl. Phys.* B **192** 332
Huang K 1982 *Quarks, Leptons and Gauge Fields* (Singapore: World Scientific) Chap. IV
Illiopoulos J and Zumino B 1974 *Nucl. Phys.* B **76** 310
Kibble T W B 1961 *J. Math. Phys.* **2** 212
Lévy M and Deser S (ed) 1978 *Recent Developments in Gravitation* (New York: Plenum)
Lopes J L 1981 *Gauge Field Theories* (Oxford: Pergamon)
Marshak R E, Riazuddin and Ryan C P 1969 *Theory of Weak Interactions* (New York: Wiley)
Naimark M A 1957 *Am. Math. Soc. Translations* Series 2 vol. **6**
Nath P 1984 in *Supersymmetry and Supergravity/Nonperturbative QCD, Lecture Notes in Physics* no 208 (Berlin: Springer)†
Nath P and Arnowitt R 1975 *Phys. Lett.* **56B** 177
Ne'eman Y, Crowin L and Sternberg S 1975 *Rev. Mod. Phys.* **47** 573
Ne'eman Y and Regge T 1978 *Rev. Nuovo Cimento* **1** no 5 (Series 3)
Neveu A and Schwarz J H 1971 *Nucl. Phys.* B **31** 86
van Nieuwenhuizen P 1981 *Phys. Rep.* **68**(4) 189†
Nilles H P 1984 *Phys. Rep.* **110C** 2
O'Raifeartaigh L 1965a *Phys. Rev. Lett.* **14** 575
—— 1965b *Phys. Rev.* **139** B1052
—— 1975 *Nucl. Phys.* B **96** 331
Ovrut B A and Wess J 1982 *Phys. Rev.* D **25** 409
Ramond P 1971 *Phys. Rev.* D **3** 2415
Rivelles V O and Taylor J G 1981 *Phys. Lett.* **104B** 131
Salam A and Strathdee J 1974a *Nucl. Phys.* B **76** 477
—— 1974b *Phys. Lett.* **51B** 353
—— 1975a *Nucl. Phys.* B **97** 293
—— 1975b *Phys. Rev.* D **11** 1521†
—— 1978 *Fortschr. Phys.* **26** 57†
Samuel S and Wess J 1983 *Nucl. Phys.* B **226** 289; B **221** 153
Santilli R M 1978 *Hadronic Journal* **1** 223
Schweber S S 1961 *Relativistic Quantum Field Theory* (Evanston, Illinois: Row, Peterson)
Sciama D W 1962 *Recent Developments in General Relativity* (New York: Pergamon)

Siegel W 1979 *Phys. Lett.* **84B** 193
—— 1980 *Phys. Lett.* **94B** 37
Sohnius M F and West P C 1981 *Phys. Lett.* **105B** 353
—— *Nucl. Phys.* B **216** 100
Srivastava P P 1973 *Nucl. Phys.* B **64** 499
—— 1975 *Lett. Nuovo Cimento* **12** 161
—— 1976a *Lett. Nuovo Cimento* **15** 588
—— 1976b *Lett. Nuovo Cimento* **17** 357
—— 1983a *Nuovo Cimento* A **75** 93
—— 1983b *Phys. Lett.* **132B** 80
—— 1984 *Phys. Lett.* **149B** 135
—— 1985 in *Relativity, Supersymmetry and Cosmology: SILARG V* (Singapore: World Scientific)
Stelle K S 1984 Lectures at the GIFT Seminar, Spain
Taylor J C 1976 *Gauge Theories of Weak Interactions* (Cambridge: Cambridge University Press)
—— 1982 *J. Phys. A: Math. Gen.* **15** 867
Volkov D V and Akulov V P 1973 *Phys. Lett.* **46B** 109
Weinberg S 1972 *Gravitation and Cosmology* (New York: Wiley)
Wess J 1983 in *Lattice Gauge Theory, Supersymmetry and Grand Unification* ed G Domokos and S Domokos (Singapore: World Scientific)
Wess J and Bagger J 1983 *Supersymmetry and Supergravity* (Princeton, NJ: Princeton University Press)†
Wess J and Zumino B 1974a *Phys. Lett.* **49B** 52
—— 1974b *Nucl. Phys.* B **70** 39
—— 1974c *Nucl. Phys.* B **78** 1
de Witt B 1984 *Supermanifolds* (Cambridge: Cambridge University Press)
Witten E 1981 *Nucl. Phys.* B **188** 513
—— 1982 *Nucl. Phys.* B **201** 250
—— 1985 *Phys. Lett.* **155B** 151
Witten E and Bagger J 1982 *Phys. Lett.* **115B** 202
Zumino B 1975 *Nucl. Phys.* B **89** 535
—— 1978 in *Recent Developments in Gravitation* ed M Lévy and S Deser (New York: Plenum)
—— 1982 *Supersymmetry and Supergravitation, Lectures at XVII Solvay Conf., LBL-15819/83*†

† Entries marked with an obelisk contain a complete list of original references.

INDEX